DECENTRALIZED PLANNING AND PARTICIPATORY RURAL DEVELOPMENT

About the Editor

Purnendu Sekhar Das (*b*. 1942) has taught Management and Economics at the Indian Institute of Technology, Kharagpur, for fourteen years (1972 to 1986) and then joined Vidyasagar University, Midnapore (W.B.) where he taught Growth Economics, Public Choice and Econometrics for sixteen years (1986 to 2002). Presently he is Visiting Professor in the Department of Industrial Engineering and Management, IIT, Kharagpur. He has edited two books and contributed many research papers in national and international journals.

DECENTRALIZED PLANNING AND PARTICIPATORY RURAL DEVELOPMENT

Edited by
Purnendu Sekhar Das
Department of Industrial Engineering & Management
Indian Institute of Technology
Kharagpur (West Bengal)

CONCEPT PUBLISHING COMPANY, NEW DELHI-110059

Cataloging in Publication Data—DK
Courtesy: D.K. Agencies (P) Ltd. <docinfo@dkagencies.com>

National Seminar on Decentralized Planning and Participatory Rural Development (2000 : Midnapore, India)
Decentralized planning and participatory rural development / edited by Purnendu Sekhar Das.
p. cm.
Papers presented at two National Seminars on Decentralized Planning and Participatory Rural Development, held at Midnapore during 30-31 March 2000 and 1-2 March 2001.
Includes bibliographical references.
Includes index.
ISBN 8180691934

1. Decentralization in government—India—Congresses. 2. Local government—India—Citizen participation—Congresses. 3. Planning—India—Citizen participation—Congresses. 4. Panchayat India—Citizen participation—Congresses. 5. Women in politics India—Congresses. I. Das, Purnendu Sekhar, 1942- II. National Seminar on Decentralized Planning and Participatory Rural Development (2001 : Midnapore, India) III. Title.

DDC 320.809 54 21

ISBN 81-8069-193-4

First Published 2005

Published and Printed by
Ashok Kumar Mittal
Concept Publishing Company
A/15-16, Commercial Block, Mohan Garden
New Delhi-110059 (INDIA)

Phones : 25351460, 25351794
Fax : 091-11-25357103
Email : publishing@conceptpub.com

Preface

Economic development of a country can be achieved through mobilization, generation and proper utilization of resources – natural, physical and human. Two principal mechanisms are used to achieve these objectives. These are: (1) price or market mechanism and (2) planning. Various combinations of these mechanisms are also tried in many countries successful implementation of market mechanism requires the existence of the conditions like perfect mobility of resources, free flow of information and knowledge, minimum government or outside angency interference and the like.

The economies, where these conditions do not exist, apply planning techniques for mobilization, generation and utilization of resources in order to increase the well-being of the people. The planning may be sectoral planning designed to achieve growth of different sectors like agriculture, industry, infrastructure, health, education etc., and/or target-group planning. While in the sectoral planning emphasis is on the production of goods and services, the target-group planning concentrates on the persons whose well-being needs to be enhanced. In the absence of free market conditions planning mechanism seeks to wipe off the sectoral imbalances and inequality in the distribution of income and wealth through judicious allocation of resources. The planning has the objective of equating the demand with the supply so that the basic needs of the people are fulfilled.

The planning is of two types: centralized planning and decentralized planning. In the centralized planning plans are made at the centre by a planning commission or planning board or planning committee. In the decentralized planning plans are made at different levels such as State, district and other lower levels. One can think of four different forms of decentralization — deconcentration, delegation, devolution and privatization. It is the devolution form of decentralization that practically empowers people to plan and to execute these plans for their own development. Devolution means transfer of power, authority, resources and autonomy to the lower level local bodies from the centre and/or the State.

In the socio-economic conditions obtaining in India immediately after independence the Indian government adopted centralized planning method as a mechanism for economic development of the country. Side by side, decentralized planning approach was also tried from early 1950s in the form of Community Development Programmes. The programmes did not yield desired results. However, being undeterred by the failure of the programme the government embarked upon the concept of Panchayats where people would directly take part in planning and executing the development programmes for their own benefit. The Panchayati Raj Institutions in its recent form were inaugurated in Rajasthan on October 2, 1959. Other States followed suit. Thus, it is seen that the decentralized planning, i.e., the planning by the people and for the people themselves, has been a very interesting exercise all over India for the last four decades or so. This Panchayati Raj System received constitutional recognition after the 73rd Constitution Amendment Act in 1992. However, it should be mentioned that very few things of the Articles such as 243G, 243H, 243Z, 243I of the Amendment are constitutionally mandatory. A great deal of liberty is left to the State Legislature's discretion. Therefore, one can not proceed on the assumption that the intentions behind the constitutional amendment will be fully reflected in the Panchayati Raj System (PRS) in the States. The financial resources of the Panchayats, whether in the form of receipts from the Government either for special purposes or for general purposes or in the form of Panchayat's own resources, and also the functioning and the performance thereof would invariably depend upon the role, functions, powers and authority that will actually be devolved on the Panchayat bodies.

Since most of the funds available to the Panchayat bodies are in the form of tied funds – tied to the schemes or projects or programmes, the role of the PRS is that of an agent working on behalf of the State or the Centre which provides it with funds.

Very few States in India have actually gone in the direction of implementing through works the intentions of the Constitution Amendment Act. Kerala, Madhya Pradesh, Karnataka and West Bengal are some of these few States which have made some progress in the implementation of the intention and spirit of the Constitution Amendment Act, but full realization of the objectives behind the Amendment Act is miles away. Even opinions about the extent of success achieved by these States widely vary from one group of researchers and analysts to the other group. Success in the case of West Bengal is found, it is contended, more in the holding of the panchayat elections at regular intervals of five years than in the actual devolution of power, authority and autonomy to the local bodies. In this respect, however, Kerala may claim to have made commendable progress in the direction of implementation of the provisions of the Amendment Act, while nothing can be said conclusively about the other States. Bureaucratic control and interference in the Panchayat functionings could hardly be done away with in any of these States. The State Legislatures of these States have enacted laws empowering the Panchayats to plan and to implement those plans, but they have rarely been implemented.

In this connection it is interesting to note that till mid-1999 the States of Arunachal Pradesh, Jammu and Kashmir, Meghalaya, Mizoram and Nagaland did not constitute State Finance Commissions (SFC).

Punjab was the first State to constitute SFC on 22 April, 1994 followed by Himachal Pradesh, Kerala, Rajasthan and Tamil Nadu which constituted their SFCs on 23 April, 1994. The SFC in West Bengal was constituted on 30 May, 1994. West Bengal SFC was the first SFC to submit its report on 27 November, 1995 about three months before Kerala SFC submitted its report on 29 February, 1996. It is also interesting to note that while most of the recommendations of the Kerala SFC have been accepted by the Kerala Legislature and put into effect, the West Bengal SFC recommendations have been accepted by the West Bengal Government only on principle; just a few of the recommendations have been accepted by enacting laws. A comparison of Kerala and West Bengal Panchayat finance is quite revealing in many respects. While in Kerala the share of own revenues at all levels of Panchayat bodies in the total of other revenues was, on an average, 37 per cent with the range between 11 per cent and 56 per cent, in West Bengal it was only 6.6 per cent with range between 2.7 per cent and 24 per cent during the period from 1990-91 to 1997-98. Own resource per panchayat body in Kerala was Rs. 5,13,750.00 per year, on an average, but it was Rs. 46,000.00 only in West Bengal during the same period of 1990-91 to 1997-98. Own resource per head in Kerala was Rs. 28.00 on an average, while it was just Rs. 3.00 in West Bengal during the same period. It is also worth noting that while in Kerala the Government grants for the Panchayat bodies have steadily and significantly increased throughout the period from Rs. 653 mn in 1990-91 to 8836 mn, in 1997-98, in West Bengal a steady rise in the government grants during 1990-91 and 1995-96 has been followed by a steady decline during the later years from Rs. 533 mn in 1995-96 to Rs. 468 mn in 1997-98. In respect of Government grants to the panchayats, in Kerala the Government grants per panchayat body was Rs. 25.00 Lakhs, on an average, during the period. It was just Rs. 13 lakhs in West Bengal, on an average, during the same period. Government grants to Panchayat bodies per head of rural population in Kerala was on an average, more than Rs. 200.00 but it was only about Rs. 80.00 only in West Bengal.

The glaring differences in the levels of performance of Panchayati Raj System may also be found between other State also. The questions that need to be addressed are: What factors explain this wide difference between the States. How far is the lack of political will, which is practically the will of the bureaucrats controlling the administration of the States, shaping, reshaping the mind-sets of the people's representatives and thus in many cases dictating their terms upon these so-called people's representatives, responsible for the slow progress of decentralized planning activities in the States? In this context several other issues also need to be addressed simultaneously. These issues relate to (1) clear demarcation of functional domain of the Panchayat bodies; (2) transfer of functions and functionaries to the local bodies; (3) integration of vertical bureaucratic functioning and horizontal panchayat functioning; (4) devolution of power, authority and responsibility to the local bodies; (5) using the services of government officer, not as arbiter, but as providers, advisers and executors; and (6) steps needed to change the mind-sets of the principal actors in the Panchayati Raj System so that the people themselves, not their representatives, think about their own well-being, plan for that and implement these plans.

To address these issues two national seminars on decentralized planning and participatory rural development were held at the Vidyasagar University, Midnapore, West Bengal on 30-31 March, 2000 and 1-2 March 2001 where several experts from various walks of life such as bureaucrats at the State and district levels, university professors, NGO workers, researchers from various parts of the country assembled, debated and deliberated upon these and several other issues. This volume contains some of the papers that were either read at the seminar or circulated and discussed to arrive at meaningful conclusions.

These seminars were held under the DRS Scheme under SAP of the UGC, New Delhi.

It is hoped that the papers presented in this volume will help the readers to get a comprehensive idea about the Panchayati Raj System in India, particularly in West Bengal – its functioning, problems, limitations and future prospects.

Purnendu Sekhar Das

Contents

List of Contributors

Bandyopadhyay, Sabari, Research Associate, DRS (UGC), Dept. of Economics with Rural Development, Vidyasagar University, Midnapore, West Bengal.

Bose, Ajit Narayan, Former Professor, Rural Development Centre, IIT, Kharagpur and Member, State Planning Board,West Bengal.

Chakraborty, Banhi, Senior Research Officer, IIT Kharagpur, West Bengal.

Chatterjee, Biswajit, Professor, Department of Economics, Jadavpur University, Kolkata, West Bengal.

Chatterjee, Tania (Majumder), Project Assistant, DRS (UGC), Dept. of Economics with Rural Development, Vidyasagar University, Midnapore, West Bengal.

Chowdhury, Jayanta, Research Officer, AERC, Viswabharati, Shantiniketan, Bolpur, West Bengal.

Das, Purnendu Sekhar, Professor, Dept. of Industrial Engineering and Management, IIT, Kharagpur, West Bengal.

Dasgupta, Tapati, Senior Research Officer, Indian Institute of Technology (IIT), Kharagpur, West Bengal.

Datta, Prabhat Kumar, Centenary Professor of Public Administration, University of Calcutta, West Bengal.

Dhara, Aparajita, University of Calcutta, West Bengal.

Ghosh, Dilip Kumar, Senior Research Officer, SIPRD, Kalyani, Nadia, West Bengal.

Hazra, Amit, Deputy Director, Agro-Economic Research Centre, (AERC), Viswabharati, Shantiniketan, Bolpur, West Bengal.

Jana, Haripada, Midnapore Planning and Development Society, Midnapore (West), West Bengal.

Maiti, Ranjit Kumar, Joint Director, State Institute of Panchayat and Rural Development, (SIPRD), Kalyani, Nadia, West Bengal.

Majumdar, Sankar, Reader, Department of Economics with Rural Development, Vidyasagar University, Midnapore, West Bengal.

Mitra, Nita, Professor, Department of Economics, University of Calcutta, West Bengal.

Nair, V.R. Pravakaran, Madras Institute of Development Studies, Chennai, Tamilnadu.

Raha, Ruchira (Sarkar), State Institute of Panchayat and Rural Development, Kalyani, Nadia, West Bengal.

Rao, M.V., District Magistrate, Midnapore (West), West Bengal.

Sethy, Sitakanta, Institute for Social and Economic Change, Bangalore, Karnataka.

Introduction

Planning is at the core of all development processes, whether that development is social, economic or cultural and whether it is at the micro-level or macro-level. Since the conditions of free flow of information and perfect mobility of resources for the proper functioning of the market economy are rarely fulfilled, planning becomes an important, though not always essential, instrument for the overall development of an economy. Planning is of two types—centralized planning and decentralized planning. While achieving higher growth rate of national income is at the heart of centralized planning, people's development forms the main plank of decentralized planning. While in the former the focus is on the national income and people is in the background, in the latter people come to the forefront; here people is at the centre and everything revolves around it.

In the decentralized planning it is the people, not their representatives or government officials, who plan for their own overall development and implement these plans and take corrective measures in the case of deviations of the actual performance from the targeted levels of performance. The idea of planning by the people for the people is not a new one. This idea took up a concrete form through the establishment of Union Boards as early as 1919, which, however, could hardly lead to empowerment of the people. After India became independent, the concept of Gram Swaraj was introduced through Panchayati Raj Institutions and a number of committees were instituted to make the system functional and effective. But not much could be achieved because of bureaucratic hold on the entire planning system and reluctance of the state governments to devolve power, responsibility and authority to the people. In the wake of the Indian Government's new economic policy of liberalization, privatization and globalization, an acute necessity was felt to activate the PRI system through devolution of power, responsibility and authority to the people.

The 73rd and 74th Constitutional Amendments aimed at breaking down this bureaucratic barriers by recognizing the Panchayati Raj Institutions as the third tier of governance. Though the PRIs received constitutional sanction, the autonomy of the system was restricted to a large extent by empowering the state legislature to enact such laws as will make these institutions function as the third tier of governance. As a result, the state legislatures remained as powerful as ever. Thus, the success or failure of the PRI system to work as an independent unit of governance largely depends on the intentions of the state government – their willingness to allow the lower tiers to emerge as the centre of power – the position the states enjoyed so long unhindered.

However, even within these constraints of otherwise well-intentioned amendments of the Constitution, some states have moved, though not always with expected alacrity, towards legislating acts and enacting laws for empowering the PRIs with necessary power, responsibility and authority, at varying degrees. West Bengal is one of those states which have achieved partial success on this front.

Against this background, the present volume underscores the importance of the PRI System, its relevance in the post-liberalization and globalization era and under WTO regime for people's, particularly women's, empowerment and development, current thinking in rural development planning techniques, self help group and participatory rural development, actual functioning of the PRIs and prospects of sustainable development in the rural areas.

In chapter 1 – Biswajit Chatterjee discusses various problems that a decentralized planning mechanism may face in a globalized regime. These problems appear in the form of informational asymmetry, micro-interventions and management of target groups and self-financing of the decentralized projects. The globalization of our agrarian economy has thrown up many new challenges but has also offered new opportunities for ameliorating poverty of the underprivileged by extricating them from the age-old shackles of bureaucratic tyranny.

In chapter 2 – People's Participation in Local Governance: The Gram Sabha Experience in India, Prabhat Datta makes a detailed study on the role of Gram Sabha which is recognized by the Indian Constitution as a statutory unit to collectively debate the issues concerning the village people. The study covers various aspects of the Gram Sabha such as its constitution, purpose and objectives and, most importantly, its functioning as culled from various reports from several states in India.

In Chapter 3 – Integration Between Physical and Financial Planning in the Decentralized Planning of Kerala: Need for a New Strategy, V. R. Pravakaran Nair focuses on the lack of integration between the district credit planning system and the decentralized planning in Kerala and the problems associated with it. To highlight the severity of the problem, he has analysed the relevant data on projects under decentralized planning for the Year 1997-98 in the Thiruvananthapuram District in Kerala. He also examines the issues related to the decentralization of credit planning and the need for an effective implementation of micro (village) credit plan which is compatible with the current decentralized planning.

Sitakanta Sethy in the paper – People's Plan for Sustainable Rural Development (Chapter 4) makes a SWOT analysis of development administration and Panchayati Raj Institutions from which he asserts that the PRIs can be made effective with the co-operation and co-ordination of both the development administration and local Non-Governmental Development Organizations (NGDOs), which, besides many other important activities, try to arouse the feeling component of the rural poor to use their brain for changing the socio-economic order in the countryside.

In Chapter 5 – Strengthening Panchayat Raj Institutions: The Case of West Bengal Dilip Kumar Ghosh discusses those articles of the 73rd Amendment of the Indian Constitution that pertain to the powers of the third tier of the government, the areas of resource mobilization by the Gram Panchayats, Panchayat Samiti and Zilla Parishads of the states in general. In this connection he makes a reference to the West Bengal Panchayat Act, 1973 (as modified subsequently) and gives a brief description of the powers bestowed upon the Panchayat bodies for augmenting their resource base at the village, block and district levels. He has made a detailed study of the trends in the assessment and collection of taxes, allocation and expenditure of fund in infra-structure building and various other important aspects of Gram Panchayats and Gram Sansads in the districts of West Bengal for different time periods.

Chapter 6 – Participatory Micro-planning for Village Eco-Development by Amit Hazra and Jayanta Chowdhury is a social assessment study of joint forest management (JFM) in the Protected Areas of Buxa Tiger Reserve of West Bengal. The paper deals with some conceptual issues related to this topic, describes some practical experiences and problems about the participatory micro-planning and develops a model for solving these problems.

In Chapter 7 – Decentralization: Learning from Midnapore, 1980-2000, Ajit Narayan Bose makes some illuminating observations on different types of decentralization and later deals with three phases of the experimentation on village planning by villagers centred on panchayats in West Bengal by the Zilla Parishad, in close collaboration with the then Rural Development Centre of IIT, Kharagpur. He discusses the lessons learnt from these experimentations on village planning during three phases – 1980-85, 1985-87, and 1990-2000, and how the concept of decentralized planning evolved over years by utilizing the experience of the previous phases of planning. He gives a brief description of how this plan was implemented by making a reference to Aurai Gram Panchayat Plan – 1990-2000 in the district of Midnapore.

The next two chapters 8 and 9 discuss the issue of women empowerment in the decentralized planning framework. The paper – Women Empowerment for Functional Efficiency in Decentralized Planning at Bottom Level by Tapati Dasgupta (Chapter 8) observes that till the early 1983, the women did not have any specific role in the making of people's government at bottom level. While the Nairobi Conference of 1985 and the 1988 National Perspective Plan for Women brought about a worldwide sensation in the women's camp, the real breakthrough in women's involvement in politics, administration and development came only in recent years with the promulgation of the 73rd Constitution Amendment of March, 1993. She wants to drive the point home, using the findings of a few case studies on the pitiable status of women in the local bodies in some states in India, that despite all tall talks about women empowerment by ensuring their involvement in politics, administration and development, the number of cases of maltreatment of the women representatives of people in different panchayat bodies far–outstrip that of rewards or acknowledgements for their activities.

Banhi Chakraborty in her paper – Decentralization of Power: A Case of Women Empowerment (Chapter 9) assesses how far the objective of empowerment of women has been achieved under the new dispensation of the Panchayati Raj Institutions. Her study is based on an empirical analysis of field survey data collected from some PRIs selected from some blocks in the district of Midnapore,

West Bengal. Using the results of her analysis, she makes some illuminating observations that the gender-biased training that starts from the women's childhood leaves little scope for them to feel that they can do any thing independently and that they are capable enough to compete with their male counterparts in many important areas of development. Another important finding of her study is that political freedom for a woman, even if it becomes a reality, means nothing if she can not exercise her rights in her own family.

In Chapter10 – M.V. Rao discusses the role of self-Help groups (SHGs) in improving the socio-economic conditions of women in the district of Midnapore. Making several case studies he has shown the extent of improvement in the economic condition of the women members of the SHGs. He has also discussed from his own experience gathered through long and the direct association with the SHGs in the district the problems facing these groups and the methods employed to solve these problems. This paper demonstrates that active participation of the administrative functionanies at the district, block and panchayat levels can transform the undeveloped rural Bengal into a prosperous one through the formation and growth of SHGs.

In Chapter 11 – Decentralized Development and Micro-credit: The Role and Efficacy of Self-help Groups – Aparajita Dhara and Nita Mitra discuss the limitations of the banking sector in issuing loans and advances and the problems of repayment of loans. In this context, they critically analyze the role of the Self-Help Groups (SHGs) in India in general and a few SHGs operating in the district of Howrah in West Bengal. Some policy issues are also discussed in the paper to make the SHG mechanism viable and effective.

Sankar Majumdar in his paper – Role of Voluntary Sector in Rural Development (Chapter 12) analyses the role of voluntary contribution of the village people in the infrastructure building in the rural areas vis-à-vis that of panchayat tax. From a sample survey of 120 families drawn from two villages under a Gram Panchayat in the district of Midnapore in West Bengal, he observes that people's voluntary contribution towards creation of facilities in their own areas is relatively much higher than the taxes that they pay the Gram Panchayats. The author suggests that if voluntary contribution replaces payment of taxes as a source of fund for development works, development will be more rapid and visible, since the fruits of development will come direct to the contributors of fund.

In Chapter 13 – Current Thinking and Experiments in Rural Development Planning in West Bengal, Ranjit Kumar Maiti describes current experiments of participatory planning process under Convergent Community Action (CCA) initiated by the Government of West Bengal in four pilot blocks of Kashipur in Purulia, Salboni in Midnapore, Mal in Jalpaiguri, and Haringhata in Nadia district.

Referring to the planning framework, methodology adopted, experiences gained and lessons learnt from action research, the author claims that sustainable development can be meaningfully operationalized by the people under this approach.

In Chapter 14 – Decentralized Planning and Rural Development: An Experiment in a Block in Paschim Medinipur District of West Bengal – Ruchira Raha (Sarkar) and Hari Pada Jana give a detailed description of decentralized planning procedure adopted under the Convergent Community Action programme at the Gram Sansad level in a block in Paschim Medinipur district in West Bengal. As the authors were directly involved in the formulation and implementation of the village plan in the block, they describe from their own experience the problems encountered, remedial measures taken and the degree of success achieved.

Purnendu S. Das, Sabari Bandyopadhyay and Tania Chatterjee (Majumder) in their paper – Panchayati Raj Institutions and Rural Development in West Bengal: A Study Based on Household Survey of Some Villages in the District of Paschim Medinipur, West Bengal (Chapter 15) – have made an assessment of the present state of socio-economic development of the village people in West Bengal on the basis of relevant data collected from a randomly selected sample of 310 households. Considering several economic and social indicators of development, they have shown that there has been no perceptible improvement in the socio-economic conditions of the village people despite the facts that the Panchayats have been entrusted with some statutory powers for making and implementing development plans and that they are allotted annually sizeable funds for this purpose.

We have traversed a long distance in time from the 'local self-government' concept of Lord Ripon to the institutions of self-government concept of today. The main objective of this tortuous journey over more than a century was to empower the people with necessary powers, responsibility and authority for their self-governance and development. For fulfilment of this objective, what is, however, urgently needed is the political will of the government to hand over power to the institutions of self-government, which, in turn, requires a harmonious integration of development administration and regulatory administration since the process of development entails a serious problem of displacement of established groups leading to social conflicts. In the international ambience of deregulation, liberalization and globalization when more and more economies are being integrated under the common philosophy of market-oriented development, the micro-administrative institutions like the Gram Panchayats under the new decentralized dispensation can really deliver the goods. More than ten years have passed since the time the 73rd and 74th Constitutional Amendments took place in India. All the states passed their acts before 24th April, 1994 in conformity with the 73rd Amendment. However, experts feel that in many cases the states have yet to fulfil the expectation of the people in giving powers to them. Holding Panchayat elections at regular intervals, reservation of one-third of seats for the women in the panchayat elections, which is yet to be passed by the Parliament, and such other things will not necessarily ensure the proper functioning of these self-governing institutions. It is not only the letter but also the spirit of Amendment that must be followed by the states, and only then the dream of 'government of the people by the people' including the poor, the disadvantaged, the low caste and the scheduled tribes, can materialize.

The papers presented in this volume make a critical assessments of the administrative and development functionings of the panchayats in some of the states in India, particularly in the state of West Bengal. Going through these papers, the readers are expected to broaden their knowledge about the state of Panchayat Raj System in India, its problems and prospects.

Purnendu Sekhar Das

1

Decentralized Planning and Economic Reforms in India

BISWAJIT CHATTERJEE

In recent years India has embarked upon new economic policies in several directions to improve the efficiency of her resource allocation, to enhance productivity and growth of labour and capital, and to make the Indian products globally competitive. While the merits or otherwise of these policy formulations and their appropriate mode of implementation have been extensively debated, there is no doubt that there has been a marked shift in the policy thinking towards more market oriented approach compared to the earlier ones of bureaucratic control and government interventions. The exact connotation of economic planning in our country is now being debated in this changed environment of economic reforms. One of the basic issues that has received attention in policy circles has been "decentralized planning" as practised at least notionally in different parts of India over last fifty years. Although the notion of decentralized planning is an old one in the literature on economic planning and on Indian economic thinking, the revival of this theme in the current policy debate deserves careful scrutiny, particularly in view of its limited experiments in some states like Karnataka and West Bengal, and prevalence of economic backwardness in different regions of our country. In this paper, we shall attempt at analyzing some analytical and practical issucs in decentralized planning in India and their implications for economic reforms in the coming days.

Planning: Centralized v. Decentralized

India has embarked on the Path of planned economic development in the overall framework of a mixed economy with both private and public sector units operating in the sphere of production and distribution of commodities. The idea has been to reap the benefits of private initiative and leadership to achieve the goal of rapid economic growth within broad contours of socially set, governmentally designed objective of social justice. Although the overall target rate of growth and the corresponding requirements of saving—investment and foreign exchange are estimated for the country as a whole by the Planning Commission, the ownership of major resources continues to be in private hands, thereby resulting in conflict of objective of the government and its planners and those of private capitalist and landlords which stand in a major way in achieving the plan targets. Following the model of Soviet planning, the so-called "commanding heights of the economy" in strategic sectors like defence, heavy industry and machinery, coal, steel, electricity, petroleum and social overhead capital like transport, communication etc., are monitored almost exclusively by the public sector, while agriculture and basic consumer goods production have been left to the private sector. There are also some areas where both the private and public sector units are operating side by side on the presumption of healthy competition and complementarity. With markets signalling what is optimal from the viewpoint of private interests, there is no guarantee in our economic system that resource allocation as per market signals and individual incentives will exactly correspond to the sectoral and temporal allocations designed in the planning exercise. As a result, we have seen proliferation of controls and regulations introduced by the government in various forms to correct the discrepancies and distortions. And in designing such controls, the planners had to exercise judgements about the inherent contradictions of interests, and choose their options from a constrained set. This problem of conflict resolution between planning vs. markets lies at the core of any meaningful planning exercise in our country, irrespective of the forms of planning, centralized or decentralized. While the theoretical as well as practical issues of the market vs. plan debate have occupied the centre stage of the contemporary discourses on political economy, it is worth noting here that unless the domains of government controls vis-à-vis the free as well as unfree markets are clearly defined, the designing of institutions of decentralized planning and their working across various tiers of economic activity are bound to be abortive in nature. This is primarily due to the fact that the local bodies often do not have access to resource on their own, nor do they have statutory powers to tap/tax all types of local resources to finance locally designed development plans for which they have to depend on funds/allocation from the state government whose supply is uncertain both in terms of timing as well as magnitude. Since the resources are primarily owned by the private individuals, and there are always some discrepancies between private and social returns on investment, the release of suitable funds for local development plans may be extremely difficult in a mixed economic environment unless specific institutions with statutory powers are created and maintained at the grassroot levels. Decentralized planning has received renewed attention in recent discussion on economic policy, because of the general belief that such a system with its new form of organizations like village panchayats and local bodies with popular participation dominating the decision-making process, would be more market-friendly and, therefore, more suited to the new policy environment of greater liberalization than the existing top-heavy bureaucratic and corruption-friendly apparatus and organizations.

Theoretically speaking, there are advantages and problems associated with either type of planning strategies and from the operational point of view, no clear-cut demarcation between centralization and decentralization is possible. Planning exercise, if it is to successfully replace the working of the market mechanism, needs not only to check the inter-sectoral and intertemporal consistency under a suitably defined optimal programmes, but also to impose macroeconomic consistency which can be decided only from a central authority. The advantages of centralization lie in having a complete control and information about the directional flow of resource and their utilization with the consequent impact on the economy as a whole. Thus, if the planners want the economy to grow at a stipulated growth path and at the same time want to maintain a given proportion between sectoral composition of outputs and resource uses, it is easier to exercise certain controls or instruments to that effect. Centralization means decision-making at the top, and if the economic system is totally coordinated and controlled, these decisions are carried out at the micro-level in the computer—like fashion without any consideration to the local needs and aspirations, which themselves may merit separate attention. The success of such exercise or programme designed centrally, however, requires the availability of not only correct signals about production, distribution and marketing at various stages of production, but also all requisite inputs and finance and organizational network at each and every remote place. Bottlenecks in any one of these factors are sufficient to upset the central plan in its entirety and in an economy where government and its agencies have to compete with private agents for the control of resources through the market, such a fully controlled experiment, as described above, is unlikely to achieve the projected targets. A lot depends on the marshalling of resources and entrepreneurship at the micro level for the success of the overall planned targets. Centralization, therefore, has to depend on decentralized activities, and the concomitant decisions, which may not always be perfectly consistent.

If, on the other hand, the focus of planning is on decentralization and micro-level project planning with all its details, the success of such a programme rests on the fulfilment of the following conditions. First, the individual economic units must have complete information not only about its resource and capability constraints, but also about the macro-variables, which determine the overall framework within which each plan has to operate. Secondly, the administrative machinery which would implement the plans drawn upon the basis of local requirements must have control over resources which they could tap to meet the plan requirements. If the local unit has to rely on the central planning authority for obtaining the resources, the success of the local plan depends upon the ability and willingness of central planners to take away resources from other centres to this centre, resulting in structural imbalances quite often. The problem, therefore, boils down to the domain of political hegemony and control, and in a country like India where we have not only a mixed economy but also a federal polity within the democratic set-up, this problem is very important in our plan formulation and its execution. Perhaps one needs to adopt a political economy approach to our planning and analyze the conflicts and tensions of various social groups or classes in our society. Such an approach will help in demystifying the belief that the establishment of Panchayati Raj system throughout the country and total decentralization of plan formulation will act as panacea to all economic ills of our country. If the power structure of the society is monocentric, then decentralized units will be defunct and useless, and in the mixed economy framework, they are likely to augment the relative dependency on the private sector. This is perhaps one reason why in our new economic policy environment of private initiatives and market signals, the enthusiasm about decentralized planning has surfaced in the foreground in recent years. It may be noted that the common feature of our new economic policy package and the constitutional amendment giving more powers and functions to panchayats and municipalities directly, has been decentralization, to the market in one case and to lower levels of governance in the other. Both the processes could turn out to be irreversible, because of the pressure of globalization in the first case, and people's power in the second. If and when the two processes find fulfilment there may be little room for central planning. Although such an eventuality has never been made explicit in the policy pronouncements so far, the compulsions of real economic forces and real politics would be difficult to be swept under the carpet through rhetoric and hurriedly drawn up relief programmes. In fact, the current tendency to announce popular relief programmes in different parts of the country to attract voters has proliferated to such an extent that the fiscal deficit of the government is threatening to assume alarming proportions. There seems to be a complete lack of co-ordination among the government sector, and centralization or decentralization of decisions cannot correct such disturbing trends, which are the results of political compulsions.

It is perhaps worth noting in this context that in India, the concern for decentralized planning is as old as planning itself, with the first plan document itself testifying the need for local level people's participation in the formulation and execution of plans, and all the subsequent plan documents echoing the same positions. The Balwant Rai Mehta team appointed in 1957 recommended constitution of statutory elected local bodies with the necessary resources, power and authority devolved on them and a decentralized administrative system working under their control, which was the basis of the Panchayati Raj system introduced in the country. In 1977, two committees, the Dantwala Committee on block-level planning and the Ashoka Mehta Committee on Panchayati Raj, have been set up by the Planning Commission and their reports submitted in 1978 gave wide-ranging suggestions for improving the functioning of village level bodies and the institutional organizations. But despite all these suggestions and recommendations, on the whole, the performance of decentralized planning experiment has been rather dismal, barring a few exceptional cases. After the initial enthusiasm in the late fifties and early sixties, which raised expectations, there has been retrogression for over two decades now. It is important to understand why the idea, though all along with us, has failed to work. Some of the reasons may be outlined below.

Since decentralized planning involves delegation of decision-making powers to the sub-state levels with corresponding devolution of resources, the performance in this respect is very much linked to the politics of planning and development within a democratic framework. In the early years of planning, the political leadership both at the central and state levels had ideological commitment towards decentralization, but in those periods the level of public expenditure was relatively small, mainly on infrastructural investments, which are by and large lumpy in nature. Over a period of time, the struggle for political power within the state level became much more acute, as the intensity of dissension within the ruling party and across regions sharpened. At the same time planning and public expenditure steadily grew in importance, providing opportunities of control and manipulation at the state and the central level and denying decentralization at the grassroots level. The allocation for schools, hospitals, roads, water works etc. which can potentially be decided upon at the district level by the elected institutions began to be increasingly manipulated by the dominant groups at the state level to their advantage by superseding the local level institutions. The fact that the local power groups could not effectively counter such actions indicates their weakness in the real politic of the country, as they were too heterogeneous to combine and express revolt against the powerful centrist lobby who wielded the money power. When poverty removal was incorporated as an explicit objective of development strategy in the 1970s, several programmes for the benefit of small and marginal farmers, landless labourers, scheduled castes and tribes were launched at the initiatives of the central government with the state governments having minor share of investments. The dominant rural elite, which was opposed to granting direct benefits to the poorest, exerted their influence at the state level to appropriate these benefits for themselves, thus making the targeted schemes ineffective. The sharpening of regional backwardness and differentiation within the rural economy in many States was due to such political forces, even within the ruling parties. As a result, it was increasingly felt by researchers that unless thorough institutional and social changes are introduced, many of these public investments would continue to remain wasteful. The constitutional amendment by the central government in 1994 was an attempt to devolve power to the local bodies directly without going through the apparatus of state government and its political managers, and at the same time bypassing the vast task of shake-up in the hierarchy of rural organizations and institutions in different parts of India. To what extent would this desire of the policy makers be successful is an open question—its success depends on the response of our polity to the advantages and disadvantages of true decentralization of power, particularly economic.

There are a number of assumptions underlying the optimism of the policy makers. It needs to be realized that statutory provisions are no doubt important and desirable, but they themselves are not sufficient to effect a break from the past. These provisions are no substitutes for the political will to delegate power to the bottom. If the elected state governments can be superseded on grounds of political exigencies, there is no basis for entertaining different expectations with regard to local level institutions. A formal delegation of power is not sufficient to ensure poverty alleviation unless concomitant measures are undertaken to bring about structural changes to weaken the socio-economic powers of the rural rich, ensure adequate representation to the poor in the local bodies through necessary electoral reforms and improve the bargaining power of the poor themselves by organizing them into activity-specific groups and through their involvement and participation in the formulation and execution of these programmes targeted for the poor. In fact, it must be understood that the centre's role in decentralized planning can at best be advisory in nature, and much of the assistance it can render towards district planning has to be channelled through the state governments, whose consent and participation becomes an essential precondition for the success of such a venture. If this simple fact is not taken into consideration, then much of the projected benefits will simply be lost in yet another battle in centre-state financial relations. Quite often, the top-heavy bureaucratic set ups in the states' administration also fail to appreciate the specific needs in local areas, and given the propensity of emitting political gimmicks in our country, they usually end up by simple adding up some mutually inconsistent and myopic plans, and release funds on purely ad-hoc basis. As a result, many of the good intentions associated with decentralized planning remained infractuous. Thus, it would be naive to imagine that much of the ills of our ill-conceived development planning would simply wither away if and only if constitutional provisions for direct devolution of power to the rural panchayats are made. In fact, if the State governments are really serious about decentralization of power to the rural poor, they need to take specific care in improving the capability of the elected representatives of the rural panchayats such that they can effectively plan according to the local needs and tap local resources to finance these projects. This will enable these bodies to be self-reliant in the true sense of the term and also ease the burden of the state exchequer. *In fact, these local bodies can be the vehicle for implementing the tax reforms agenda in the rural sector like the agricultural holding and income tax, which can tackle the problem of tax compliance for which such tax measures could not be implemented for such long period in India, and the rural rich who have benefited so long from all rural development schemes without any contribution to the exchequer, can be brought to the mainstream of tax administration.* If such schemes of tapping rural resources become viable, the other resources which needs be channelled towards rural development may be utilized to finance big public investment on infrastructure and human resource development.

Economic Reforms

We now come to the issue of economic reforms. In fact, after about a decade of economic reform experiments in India since July 1991, now we are talking about the need to initiate the second generation of economic reforms in our country. For all practical purposes, these measures are supposed to include the unfinished agenda of economic reforms. But on a philosophical plane, this would involve redefining the contours of the state in our economy. From the prevailing centrist tendency to usurp power, and distributing to the states only to the extent compulsory on political grounds, the need to empower the bottom layers of our population has come into the sharp focus in our reform agenda in view of the enormous gap that were discerned during the plan period between the stated objectives of our centralized planning and the actual accomplishments in development programmes. There has also been a concern that unless the participation of the targeted beneficiaries in the development programmes is ensured, the distribution of the fruits of development to the bottom layers remain unfinished. Moreover, we have entered into a phase of reforms in our liberalization programme, where growth, efficiency and effective international competitiveness of not only the numerous small-scale and household industries spread over the corners of the country are to be improved, but also of a variety of agricultural commodities, which require the involvement of farmers as well as poor agricultural labourers on a large scale. Decentralization of economic power to the grassroot level, therefore, becomes essential not only for the sake of ensuring people's democracy, but also for ensuring higher growth and equity. Our participation with the emerging liberal global trading arrangements as defined by the WTO is now a foregone conclusion, and effective steps are to be taken such that transition to the new regime can bring gains to the economy and society in general, and the hitherto deprived in particular. Decentralized planning and associated devolution of power to the village panchayats is new institutional arrangements, which can be effectively utilised to meet the requirements of the economy in the emerging environment. The 73rd and the 74th amendments to the Indian Constitution have already initiated the process of administrative decentralization to the gram panchayats. They are now to be implemented by the state governments to effectively involve the people with the task of rural development, and unleash a process of their empowerment, which has long been ignored in our march towards socialistic pattern of society. In fact, with the onset of the inevitable process of globalisation into the Indian Economy, the task of people's participation and empowerment through the institutions of decentralised planning becomes extremely important in managing the growth process in our rural economy. Briefly, the task may be described as follows.

Challenges Ahead

The agricultural sector constitutes the largest safety net in the Indian economy, as it absorbs the largest workforce of our economy in both farm and non-farm occupations, and its good performance also ensures food security to our large population. One of the major constraints faced by this sector since the middle of 1980s has been that capital formation in this sector has declined in real terms, and the total factor productivity growth in this sector has gone down to less than 2 per cent per annum, which seriously constrain its further growth potentials. Given India's commitment to the principles laid down by the WTO, the globalisation of the agricultural sector in our country seems inevitable. The immediate impact of such a switch in policy regime would be an upward trend in agricultural prices including food, which may tend to augment the incidence of rural poverty. But there is an income effect of this change of regime, with implications for domestic income distribution and hence the effective demand for the non-agricultural sector, mainly the industrial sector. We need to enhance agricultural growth not to ensure greater food supplies in the country (in fact, per capita food availability at present exceeds the average per capita nutritional requirements of our population, although there are instances of distribution failures), but to ensure higher purchasing power for both food and non-food items through greater employment and income in our agricultural sector. With scope for employment shrinking in the short and medium run as a consequence of globalisation induced labour displacement in the organized industrial sector, and in the labour saving service sector (although some segments of it might have higher employment elasticity), a reasonable degree of social safety net could be provided through additional employment generation in the agricultural sector. This would be possible only with higher doses of investment in this sector so as to relax the supply constraint mentioned above. The main problem today is to generate funds required for financing such investment on a large scale (small doses will not help and in fact would wash away), and for this, the hidden surpluses in the rural economy need be tapped by suitable design of incentive cum taxation policies that are politically implementable. This would require a decentralized administrative mechanism where participation by the local people would monitor the mobilization of local resources and their uses, including marketing of the products such that the real gains of globalisation induced expansion in agricultural activity accrue to the tillers of the soil and the real wage rate of the agricultural labourers could be augmented so as to reduce the degree of rural inequality. Secondly, a decentralized rural financial institution with people's representatives and shareholding by landlords and peasants at the same time and group insurance towards checking of default could be developed with the help of elected panchayats to meet the problem created by the imperfections in the rural credit markets, which often binds the poor agricultural labourers into different forms of exploitative interlinked contract with the rich landlords. These institutional reforms, as suggested above, if effectively carried out and targeted, would ease out the pressure given by the upsurge in poverty parameters in the post reform period, and would make the task of poverty alleviation much more directed and specified than are presently. There is no reason to protect the rich in both rural and urban areas from the onslaught of globalisation, rather they should be induced to participate in it and pay for the reforms obtainable from the expansion of economic activities, which in turn, may be used to provide safeguards to the ultra-poor in our society. For this, we need a careful planning and monitoring of the poverty alleviation schemes spread across states by a central expert agency endowed with detailed information and targeting them properly with the help of the rural institutions like panchayats such that the beneficiaries can directly monitor the successful implementation of the schemes, and the wastage could be plugged. The additional resources mobilized from the hitherto beneficiaries of rural development must be efficiently utilized not only to reduce poverty, but also create social infrastructure in the rural areas so as to augment agricultural productivity and diversify the production base in the rural sector of our economy. But all these would require careful management of the rural institutions of planning, finance and targeting of our poverty administration, which must be developed in different States in the coming years to meet the new challenges of globalisation of Indian agriculture.

Conclusion

In conclusion, we may note the following general principles/issues that lie at the heart of the controversy regarding decentralized planning and economic reforms in a country like India. First, the framework of general equilibrium analysis and the functioning of atomistic markets are supposed to be mimicked by a decentralized planning mechanism of Lange-Lerner-Malinvaud variety, but informational asymmetry can pose serious problems for decentralization to deliver efficiency as per welfare norms. Secondly, micro-interventions and management of target groups are the twin major challenges of decentralized planning. Apart from the organizational network which evolves under the patronage of the state government and the ruling political party, as in the case of West Bengal, the issue of self-financing of the decentralized projects chosen at the local level has been of paramount importance in policy discussions. The problem gets magnified because of the overlaps and externalities between local units and the delicate nature of political pressures that such a financing scheme implies in a federal democratic polity like India. The constitutional amendments have opened up new possibilities of defining the implementable set of schemes which can be self-supporting as well as surplus-generating, which are urgently needed to sustain the growth process of our rural economy in a globalised environment. Thirdly, while it is true that the financial sector often reflects many developments in the real sector imbalances, there also exist cases of indirect effect of financial sector variables on the growth of the real sector, and the institutions developed by the state need be examined in this context. If these institutions are financial in nature, there is a ticklish problem of loan recovery which is linked with the viability of these financial institutions. Alternative forms of decentralized management of the rural financial institutions based on share-holding and risk-sharing by the landlords and peasants, and aided by the elected panchayats need to be explored with a critically assigned role of the state as the provider of seed capital, and this may break the inter-linkage between credit and labour markets in our rural economy, and pave the way for real-growth induced productive accumulation of surplus in our rural economy, particularly after the full-scale liberalization of the agricultural sector.

A lot has been discussed in the literature on the merits of decentralized planning, particularly in view of the apparent success stories of West Bengal, Karnataka, and Kerala. The lessons of these experiences definitely guide the policy planners in framing the designs of decentralized planning and grass-root democracy for other states of India in the days to come. But in respect of real empowerment of the poor people, a lot remains to be accomplished, as the principles of shared growth are yet to be translated into practice, rhetoric galore notwithstanding. The new challenges of globalisation of our agrarian economy pose not only new compulsions of making the growth process self-sustaining and recombinant, but also new opportunities of decentralized decision making by the participants of the growth process in different forms to help to ameliorate the poverty of the under-privileged without the tyranny of the inefficient bureaucracy raj in the name of management of poverty alleviation programmes in our rural economy. In this paper, we have laid bare the contours of such a strategy of decentralization under a globalized regime.

REFERENCES

Chatterjee, Biswajit (1988): *Decentralized Planning In India—Conceptual Issues And Policy Options,* IEA Conference Volume 1988, Calcutta.

Chatterjee, Biswajit and Kundu Amit, (1999a); Efficiency Wage, Employment, Technology and National Welfare in an Open Economy, *The Indian Journal of Labour Economics,* Vol. 42, No. 4, 1999.

Chatterjee, Biswajit and Kundu, Amit (1999b): Monitoring Savings and Rural Financial Institutions of India—A Rethinking, paper presented at the national conference of CESP, Jawaharlal Nehru University, October 1999.

Mukherjee, Nirmal and Bandyopadhyay, D. (1993): *New Horizons for West Bengal's Panchayats*—A Report for the Government of West Bengal, February 1993.

Rao, C.H. Hanumantha (1989): Decentralized Planning: An Overivew of Experience and Prospects, *Economic and Political Weekly,* February, 1989.

2

People's Participation in Local Governance

The Gram Sabha Experience in India

PRABHAT KUMAR DATTA

The concern for participation led development planners, experts and academicians to conceptually elaborate the meaning of participation and its various connotations. Two broad approaches can be distinguished, one of which is found in the works of the proponents of decision-making approach to development. For instance, Karl Deutsch, while viewing participation as a psychosocial phenomenon, underscored its importance in the planning process in development. For Deutsch participation denotes a technique for setting goals, choosing priorities and deciding what resources are to commit to goal attainment.

There is another approach, which minimise planning and stresses the individualistic participation of the producer, the marketer and the consumer. This strategy has been advocated by scholars like Hageboeck, Allen and by some international agencies like the World Bank, United States of America International Development. This strategy suggests minimising public sector, dismantling controls of all kinds and restricting administration to providing help to the private sector. This approach has been criticised for its neglect of basic human need and overemphasising the increased production.

Participation has, for the last 20 years, been widely used in the discourse of development. For much of this period the concept has referred to participation in the social arena, in the community, in the development projects. This led to the rise of the concept of participatory development as a means of strengthening the relevance, quality and sustainability of the development projects. Of late, participation is viewed in the context of good governance which is generally understood as "both a broad reform strategy, and a particular set of initiatives to strengthen the institutions of civil society with the objective of making government more accountable, more open and transparent and more democratic".

Local governance calls for an increased participation of the civil society in activities that traditionally formed part of the public sphere. It is argued that it will improve the efficiency of public services, that it will make local government more accountable and transparent, and it will deepen democracy—complementing representative forms with participatory forms.

One of the most popular forms that has opened spaces for a wider and deeper participation of citizens at the local level has been the decentralisation process which has got unprecedented momentum in 1990s. Five million people in all but 12 of the 75 developing countries have implemented some form of decentralisation with varying degree of financial and political power. Parallel to these developments, enabling legal frameworks and institutional channels for citizen participation at the local level have been developed in many of these countries, as shown below:

Legal Enabling Environment for People's Participation

Philippines : Local Government Code (1990)
India : 73rd and 74th Constitutional Amendments 1993-94
Honduras : Municipal Law (1990)
Bolivia : Popular participation Law (1992)
Namibia : Local Authority Act (1992)
Uganda : Local Government Act (1997)
Tanzania : Local Authorities Act (1992)

Selected Example of Institutional Channels for People's Participation

Tanzania : Ward Development Committee
Zimbabwe : Village Development Committees
Uganda : Resistance Councils and Committee
Nepal : Village Development Council
Colombia : Overseeing Committees
Bolivia : Village Committees
India : Gram Sabha and Ward Committees

The 73rd constitutional Amendment in India has thrown up many new challenges to the rural local government one of which is grassroot participation. The amendment has attempted to institutionalise people's participation through a body called Gram Sabha, which had earlier existed in some states more in name than in action. While Gram Sabha has found a place in all conformity legislation, the State like West Bengal has taken a step forward by creating a new body called Gram Sansad at each electoral constituency level. This paper seeks to examine working of the Gram Sabha in India and Gram Sansad in West Bengal. It also offers suggestions about how to improve the working of these institutions.

Gram Sabha literally means village meeting—a meeting of the villagers to discuss the common problems they face in their village. In many ways, it is comparable to the citizen forum of direct democracy in the ancient Greek City states.

Gram Sabha is body corporate while the Gram Sansad is a diffuse group of voluntary individual. The relationship between them is not one between the ruled and the rulers, but the one between the electors and the elected, the former is accountable to the latter.

Gram Sabha is an institution in the hands of the people to air their grievances and to watch the functioning of the panchayats. It also offers the panchayats an opportunity to share their difficulties with the people who have elected them. This is more necessary in a situation of rising expectations and limitation of funds. It can go a long way in legitimising the authority of the panchayats and ensuring transparency of their activities, inculcating in the people respect for political institutions and the political leadership. It can in the real sense be the guide, friend and philosopher of the gram panchayat.

Gram sabha can be an effective forum for mobilisation of the masses for community activities by enabling the people to get together under one umbrella and to interact on issues of mutual interest. In this way, it can work as a bulwark against casteism and communalism.

In India, Gram Sabha has long been recognised as the lowest unit of the grassroots democracy. Jayprakash had observed that the Gram Sabha signified the village democracy and that the relationship between the panchayat and the Gram Sabha should be like that between the cabinet and the assembly. The Gram Sabha had found a place in State level legislation during the phase of the first generation of panchayats. For example, there was the provision of the Gram Sabha in the Bombay Village Panchayat Act of 1959. The Act provided for a minimum of two meetings of the Gram Sabha to be held in a year. It was obligatory on the part of the panchayat to place before the Gram Sabha the annual statement of accounts, the report of the administration of the preceding year, the development and other programmes of work proposed for the year under consideration, the audit notes and replies thereto. The Maharastra Land Reforms Review Committee had recognised it even in the vital matters like the implementation of the Tenancy Act. But the Gram Sabha did not function well. The Diwakar Team observed in 1963 that the Gram Sabha meetings were thinly attended and a quorum was seldom achieved.

The Asok Mehta Committee had given birth to the second-generation panchayats. The idea of Gram Sabha was accepted by some states. For example, the Karnataka Panchayat Legislation of 1985 had provided for Gram Sabha. It was endowed with a lot of functions and was required to meet from time to time but six months shall not intervene between any two meetings. But the meetings were not regularly held. From the study conducted by Joshi[1] it appears that the gap between the first meeting of the Gram Sabha held at Badaga Yekkar and the second meeting definitely exceeded six months. A study of working of Gram Sabha in seven villages of the Mangalore taluk during the year 1990-92[2] showed that the Gram Sabha could be held even when seven persons were present. One of the minutes of the Gram Sabha held at Delanthabettu on August 28, 1987 stated that the members should persuade the absentee members to attend the meetings of the Gram Sabha at least in the subsequent period. Surprisingly, in the subsequent meetings the number of members present was not even half the number present in the previous meeting. On December 26, 1989 a meeting of the Gram Sabha was held at Kandemula. Only 12 members graced the occasion. The village level workers and the officers concerned were not present to listen to the grievances of the people. In the meeting of the Gram Sabha held at Permuda it was felt that the various problems of the village could not be discussed because the government officers concerned were not present.

Gram Sabhas in the Seventy Third Amendment Act

The Seventy-Third Amendment of the Constitution recognises Gram Sabha as a statutory unit. It is defined as a body of persons registered in the electoral rolls relating to a village within the area of a panchayat. The gram sabha is the reservoir of people's power, an embodiment of

Gram Swaraj. It is the place where the panchayats and the people have to stand in a face to face position. Gram Sabha brings the village community under one roof to collectively debate the issues concerning the gram.

Gram Sabha in the State Legislations

The Gram Sabha has been incorporated by all the States in their amended laws on panchayats, although there are variations in respect of numbers of meetings to be held, the issues to be discussed in the meeting and the like. West Bengal has, however, taken a step forward by making provision for Gram Sansad apart from Gram Sabha. All the voters of each of the electoral constituencies of a Gram Panchayat constitute a Gram Sansad and all voters of a Gram Panchayat area constitute a Gram Sabha. A Gram Sansad is required to guide and advise the gram panchayat in regard to the schemes for economic development and social justice undertaken or proposed to be undertaken in its area and may—

(a) identify or lay down principles for identification of the schemes which are required to be taken up on priority basis for economic development of the village;
(b) identify or lay down principles for identification of the beneficiaries for various poverty alleviation programmes;
(c) constitute one or more beneficiary committees consisting of not more than five persons who are not members of the Gram Panchayat;
(d) mobilise mass participation for community welfare programmes and programmes for adult education, family welfare and child welfare; and
(e) promote solidarity and harmony among all sections of the people irrespective of religion, faith, caste, creed or race and records its objection to any action of the pradhan or any other member of the Gram Panchayat for failure to implement any other development scheme properly or without active participation of the people of the area.

The Gram Sansad meets twice in a year. The half-yearly meeting to be held in November considers the budget for the coming year and the written opinion of the members on the budget and the last audit report. The annual meeting of the sansad to be held ordinarily in May considers the revised budget of the previous year, the accounts of the last six months, the activities undertaken last year and proposed to be undertaken in the coming year. The resolution adopted at the meeting has to be read out at the meeting itself. The meetings of the Gram Sabha and the Gram Sansad are to be convened by the Pradhan or in his absence, by the Upa-Pradhan. The inability on their part to convene the meeting empowers the state government to remove them. If the panchayat is held responsible, the state government can dissolve the gram panchayat concerned.

The Gram Sabha meets ordinarily in December every year after the completion of the half yearly meeting of the sansads in November. The agenda items of the sansad are included in the Gram Sabha meeting. It is the duty of the gram panchayat to place before the meeting of the Sabha the budget of the Gram Panchayat, annual plan, the last audit report, and the income and expenditure of the last year. The panchayat has also to place before the meeting of the sabha the decisions of the Gram Sansads along with the observations of the Gram Panchayat. The opinions of the sabha have to be recorded in their solution book. The resolutions of the sansad and the sabha are considered by the panchayat concerned the failure of which is recorded in the audit report as a case of gross irregulairty.[3]

As in West Bengal, the Gram sabha in almost all the States meet twice a year. In Assam, Bihar and Madhya Pradesh the Gram Sabha meets four times a year, while in Tamil Nadu, thrice in a year. In Madhya Pradesh the dates of the Gram Sabha meetings are fixed by the executive order of the State government and these dates are April 14, August 20, October 2 and January 26.

In most States, the functional domain of the Gram Sabha is limited to discussions of annual statements of accounts, administration report, selection of beneficiaries of the anti-poverty programmes etc. Only a few States. [Haryana, Punjab, Tamil Nadu] give power of approval of budget etc. to the Gram Sabha. West Bengal has, however, given extensive powers to the Gram Sansad and Gram Sabha.

Generally, there is a quorum for the meeting of the sabha. For example, it is one tenth of the total membership in Assam, Haryana, Sikkim and Orissa; one third in Tamil Nadu and not less than fifty in Kerala. No Act makes mention of the presence of the women. There is no mention of an adjourned meetings in the Tamil Nadu Act and in the rules of the Orissa Act. In several cases, only a reduced quorum is required for adjourned meetings. This may help the unscrupulous elements to convert the gram sabha into their private clubs.

Reports from the States

Reports from the States indicate that the gram Sabha meetings are not being held regularly. The Institute of Social Science team had found in a village in Madhya Pradesh that by December 1995, three meetings were held as against the legal requirement of six meetings. In this respect the record of West Bengal in not discouraging. In 1995, 63 per cent of the November meetings of the gram sansad were held. The figure rose to 88 per cent in 1996. However, a situation where the people have to move to the court for compelling the State Government to hold panchayat elections and where the ruling party or parties keep on postponing elections to panchayats on one pretext or another, is not absolutely unexpected.

The MP study shows that majority of them did not attend meetings because the people felt that nothing happened at such meetings. The villagers pointed out that the Gram Sabha had passed a resolution to construct the school building replacing the one-roomed tin-roofed structure. But the resolution could not be implemented since the sarpanch replied that he had no money for that purpose.

The study had further shown that the attendance was thin because the villagers did not know about the dates of the meeting fixed by the government. Even the sarpanch did not know when the next meeting was due.

In some of the States the Gram Sabha meetings have also started yielding desirable results. The Gram Sabha meeting in Karnataka successfully combated the time-honoured Devdasi system,[4] which prevailed in 167 villages of Belgaun district. The UMA research team of Bangalore[5] had witnessed a Gram Sabha meeting at Indoor gram panchayat in Uttar Kanada district. The sabha witnessed uproarious scene when the people demanded an explanation from the secretary about the activities. Only a handful of the members were familiar with the Act. The presence of women members was very insignificant. Those who came were passive observers. Many were hesitant to confront the elected representatives. Most of the questions were raised by the youth. The elders had hardly opened their lips.

The Participatory Research in Asia team[6] has studied the working of the Gram Sabha in Gujarat, Haryana, Himachal Pradesh, Kerala and Uttar Pradesh. The team has noticed that although meetings are being held almost regularly, quorum is hardly achieved. And surprisingly, despite the lack of quorum, the proceedings are prepared.

The team is generally of the opinion that there is lack of awareness on the part of the members not only about the time and venue of the meetings but also about the usefulness of such meetings. While talking to the members present in the meeting, the researchers felt that many of them were confused about the role of the Gram Sabha. Some of them perceive the Gram Sabha meetings as the political meetings and the only function of the Gram Sabha is to prepare the list of beneficiaries under the different anti-poverty programmes. They have observed that the results have been encouraging where the heads of the panchayats or the members have interacted with the people occasionally.

The team points out that, except in Kerala and Madhya Pradesh, the advice and suggestions of the Gram Sabha are not even binding on the gram panchayats. In reality, Gram Sabha meetings do not really serve the purpose of either establishing village priorities and over-all policy frameworks for gram panchayat or regular monitoring the use of the resources and implementing the decisions of the gram panchayats.

It is also observed that most of the heads of the panchayats lack skills in conducting meetings in a manner that could promote wider participation of the people in the discussion of the issues affecting them.

Issues of social justice and internal dynamics of the village have rarely been agenda items in the Gram Sabha meetings.

In West Bengal Gram Sabha created by the amendment of the West Bengal Panchayat Act, 1992 has started working well. Webster's research had demonstrated that apart from the compulsory meetings of the Gram Sabha, the informal discussions in some localities and regular public meetings were other forms of participation. Lieten had remarked that although the *boktrita* culture was not altogether absent in the Gram Sabha meetings, participation used to take place. Around one fourth of the males he spoke to had stated that they had participated in the meeting[7].

In West Bengal, the people of all age groups attend the gram sansad meetings. In most of the meeting, the panchayats publish the reports. There had been many occasions when the meeting generated a lot of heat. The author himself witnessed a situation where a group of members walked out when the pradhan refused to concede their demand.

There is still a lack of awareness on the part of the pradhan and the members about the legal provisions relating to the gram sansad and gram sabha meetings. The official present in the meetings have to correct them.

The studies done by the present author in West Bengal[8] indicate that the meetings have been more successful in places where the gram panchayats and the parties have taken initiative in synchronising them with the cultural programmes in tune with the local traditions. The modern institutional assemblies need to be tuned with the natural rhythm of the rural life.

Concluding Observations

Building institutions at the doorsteps of the people alone cannot ensure the participation of the people at large. The objective of ascertaining the wishes of the people on matters pertaining to their interests is highly commendable, but the question is: whose wishes? There is no dearth of studies to prove that the rich in the society dominate the panchayats.

The poor people do not think that panchayats take care of their interests. They get alienated. Participation to be meaningful and genuine calls for the creation of certain objective conditions in the society and motivation in the participants. The delicate democratic plant, as both the liberal and the Marxist scholars have agreed, cannot grow in an atmosphere of inequality and poverty. This is not enough. Participation in a meeting calls for education and skill. Education has, therefore, to be universalised.

To make participation of the people more effective, the following specific measures, apart from the general measures suggested above, may yield desired results. In the first place, the political parties, as in West Bengal, should come forward and play effective role in educating the people about the usefulness of panchayat system and encouraging them to participate in Gram Sabha meetings.[9] They should move from door to door before the meetings as they do before the election. Second, in the meetings the panchayat leadership should demonstrate patience and tolerance. They should listen to the constructive criticism of the people. Chandrasekhar and Inanbathan study has shown that the intolerance led to the cancellation of the Gram Sabha meetings in the past. Third, an atmosphere has to be created where the people would feel free to speak about their problems. Fourth, sustained information dissemination of positive examples of the functioning of the Gram Sabha and Gram Panchayat needs to be supported by the government at the State and national levels in order to create positive climate for promoting opportunities for self governance at the village level. The press should focus such meetings. Fifth, the government officials should be so oriented as to be responsive to the people. Sixth, the advice and decisions of the Gram Sabha should be made binding on the Gram panchayats. Seventh, as in West Bengal, the State Acts should make penal provision for the recalcitrant panchayats which fail to hold meetings of the Gram Sabha.[10] Eighth, as the villagers are not used to participating in a situation like this and the village is not tuned to it, the synchronisation of the meeting with the cultural programmes in tune with the rural life might be helpful. The last, but not least, is the generation of motivation among the members. This can be attempted in two ways, first, by making them realise the usefulness of this meeting and second, by ensuring that the decisions of the meeting are carried out.

NOTES AND REFERENCES

1. G.V. Joshi (1995), Gram Sabha in Karnataka: A non-starter? *Kurukshetra,* April.
2. M.A. Oommen (1996), *Panchayati Raj Development Report,* ISS, New Delhi, p. 27.
3. Ibid p. 28.
4. Department of Panchayats and Rural Development, Government of West Bengal.
5. P. Menon (1996) Sex slaves no more: Devdasis of Belgaum District Break Free, *Frontline,* February 9.
6. See, Uma Prachar (1997) January-March.
7. Participatory Research in Asia (1997) Proceedings of the Seminar (unpublished) Strengthening Panchayati Raj Institutions in India held at India International centre, New Delhi on August 30, 1997.
8. N. Webster (1992) *Panchayati Raj and Decentralisation of Development Planning in West Bengal,* Kolkata: K.B. Bagchi, p. 99.
9. G.K. Lieten (1994) Rural Development in West Bengal: Views From Below, *Journal of Contemporary Asia* Vol. 24 No. 4 p. 328.
10. P. Dutta (1996) 'Politicising Rural Development: Lessons From India's West Bengal", *Asia-Pacific Journal for Rural Development* December 1997 and (1997) Politics at Blade and Grassroot, *The Telegraph,* July 10.

3

Integration Between Physical and Financial Planning in the Decentralized Planning of Kerala

Need for a New Strategy

V.R. PRAVAKARAN NAIR

Introduction

During thc last few years, the Kerala economy has witnessed far reaching changes in Government policy towards the real sector under decentralized planning. The decentralized planning has at least established a system at the village level through which the villagers themselves harness such scattered information and knowledge in evolving future development activities. The decentralized planning in Kerala gives emphasis on keeping the quality of services, especially of those supplied by the public sector. In the field of real sector changes have occurred with respect to land and water management, development of agriculture and allied sectors, rural and cottage industries, drinking water, sanitation and many of the social services. The decentralized planning in Kerala envisages overall development, agricultural growth and poverty alleviation through appropriate projects. But for that, it needs a substantial amount of resources from financial institutions. The existing credit planning system cannot take into account the specificities at the regional and micro levels. Local level units are unable to use the district credit planning exercise for understanding and preparing plans to exploit the potentials in their own service areas. The decentralization of credit planning through village credit planning system could be a better and well-orchestrated option for the effective grass root level implementation of development projects under decentralized planning in Kerala.

It is argued that there is a lack of integration between the district credit planning system and the decentralized planning in Kerala. This can be evident from the fact that for the past few years, the contribution of financial institutions to the decentralized planning was not satisfactory. For example, the expected resource mobilization from financial institutions, on June 30th 1998 was 10 per cent. But the realized amount was only 95 per cent. A disturbing trend has been the reduction in the share of loans from financial institutions from 10.58 per cent in 1997-98 to 7.35 per cent in 1998-99. As a result, most of the local bodies had to revise their plans as they failed to mobilize the targeted level of resources from financial institutions during the previous year. In short, there has to be a proper integration between the rural credit planning and decentralized planning at the sectoral, regional and micro levels which is conductive to achieving the contemplated rate, pattern and composition of growth and other distributional objectives of development under decentralized planning.

The introduction of decentralized planning in the real sector in Kerala necessitates the decentralization of credit planning to mobilize additional financial resources for the successful implementation of the development plans.

There is an argument that, in the current decentralized planning process in Kerala, special emphasis is being given to those projects that would meet the criteria laid down by the financial institutions. Further, the transparent and participatory process of project formulation of the people's campaign would put sufficient peer group pressure on the borrowers for compliance to the repayment commitment. Thus, it is argued that the decentralized planning presents a unique opportunity for the financial institutions to enter into a partnership with local level planning to accelerate economic growth in the state without compromising their managerial functions. However, a major challenge of decentralized planning in Kerala was the lack of possibility for tapping additional resources from financial institutions. Credit from financial institutions forms the basis of financial resources. Thus, there is an urgent need for the integration of credit plans of the banks and the development plans of the local bodies.

The present paper seeks to define and discuss the problem of lack of integration between financial planning and physical under decentralized planning in Kerala. For this purpose, the data on projects under decentralized planning for the year 1997-98 in the

Thiruvananthapuram district is analysed to understand the expected and realised amount of credit. This will help to understand the severity of the lack of credit problem under decentralized planning. After analyzing the lack of credit linkages under decentralized planning, we try to examine the issues related to the decentralization of credit planning and the need for an effective implementation of micro (village) credit plan which is compatible with the current decentralized planning.[1]

Credit Planning in the Context of Decentralized Planning in Kerala

The decentralized planning envisages the rural development by improving the living standards of low income population residing in the rural areas and keeping the process of growth self-sustaining. Credit is a powerful instrument in bringing about rural development and agricultural growth. This is all the more true for the poorer sections in the economy as their saving capacity is constrained by low level of income. They are living in a vicious circle of poverty with low income, low savings, low capital, low productivity and, consequently, low income. A systematic approach to a credit planning process is very essential in ensuring access to credit in the rural areas for increasing agricultural production and alleviating poverty. Thus, the efficiency of rural credit delivery system has been an area of attention in the decentralized planning process at the local level in Kerala.

The micro-level credit planning simply means the identification of relevant socio-economic data for a particular region and their transformation into a development plan. Ms. Marguerite Robinson of Harvard Institute for International Development states, "micro finance refers to small scale financial services provided to people who farm or fish or herd; who operate small or micro enterprise where goods are produced, repaired or traded; who provide services; who work for wages or commissions; who gain income from renting out small amounts of land, vehicles, draft animals or machinery and tools to other individuals and groups at the local level of developing countries, both rural and urban."[2] This definition emphasises the development of priority sectors like agriculture, artisans, small scale industries. Scheduled Castes and Scheduled Tribes. However, as against this, in Kerala, the credit planning system still remains essentially moored to the financing of large and medium industry and trade.

The existing system of credit planning is called the Potential Linked Credit Plan (PLP)[3] prepared by NABARD for each and every district. The basic objective of the Potential Linked Credit Plan is to map the existing potentials for development and evolve a mechanism for their exploitation in a specified time frame. These plans are based on analysis of field data as well as secondary information available with different agencies including development departments of the state government. The Potential Linked Credit Plan estimates are then used for the preparation of back ground paper to help branch managers in the preparation of their own Service Area Plans (SAPs). The individual banks are formulating their Service Area Plans (SAPs) according to the Potential Linked Credit Plan prepared by the NABARD. The individual banks are not assessing the potentials of their own service area by themselves. Thus, the credit planning constitutes a process from above. This goes against the decentralized planning process where the priorities for the rural development are determined at the village level. The response of the existing system of credit planning to the local level development planning is to be judged by its contribution to achieve its development plans.

Credit Linkages for the Projects in the Year 1997-98 Under Decentralized Planning in Kerala

The following section analyses the pattern of resource mobilization for the projects in the year 1997-98 in Thiruvananthapuram district under decentralized planning. It also examines the expected and realized amount the credit from financial institutions.

The Thiruvananthapuram district consists of 84 gram panchayats, 4 municipalities and 1 municipal corporation beside the district panchayat.

The decentralized planning in Kerala focuses on projects drawn up locally and not implementing schemes handed down from above. Even when schemes are undertaken as part of people's planning, specific projects would have to be prepared as part of the local plan before they are implemented (Issac, 1999).

We can make a distinction between a project and a scheme. Schemes are development activities that are drawn up from above and implemented uniformly in all localities as per the norms and guidelines of the scheme. For example, building a road within the rural employment scheme is a project. The road takes into consideration local conditions. It is a small unit of the scheme, and should be implemented within a limited time period. Local projects can be part of a scheme or they can be independent of any scheme.

The projects in different sectors may be classified into three:

1. Projects in the productive sectors;
2. Infrastructural projects;
3. Service and welfare projects.

An important part of a project document is the allocation of funds for its successful implementation. Thus, financial analysis is important to determine the amount of investment or the money that will have to be invested to implement the project. The financing pattern of a particular project should be analysed to understand the availability of the funds required for the implementation of the project.

In the following paragraphs we analyze the financing pattern of projects of local bodies in the Thiruvananthapuram district and the importance of credit linkages in these projects for the year 1997-98.

Table 3.1 provides the details of the total outlay of the projects of different local bodies and its division into major sources. It also gives the percentage distribution of the total outlay expected for all the local bodics. The total outlay for all the local bodies for the year 1997-98 was Rs. 173.87 crores. However, the grant-in-aid from the state government constituted only Rs. 77.13 crores i.e. 44.36 per cent. Thus, in order to realize the total expenditure the projects under decentralized planning had to mobilize additional local resources of Rs. 96.74 crores.

From the Table 3.2 it can be seen that the second larger contribution was in terms of beneficiary contribution, i.e. Rs. 39.71 crores, followed by loans from financial institutions of Rs. 22.40 crores. The percentage shares of beneficiary contribution and financial institutions were 22.84 per cent and 12.89 per cent respectively.

Three sources of fund show almost equal amounts of contribution to the total outlay. They are own fund of Rs. 7.44 crores, state sponsored schemes of Rs. 7.78 crores and 8.35 crores of voluntary contribution. The contribution by centrally sponsored schemes and co-operatives were Rs. 5.88 crores and Rs. 2.56 crores respectively. It is important to note that avoiding state and centrally sponsored programmes, the expected contributions made by the local bodies in terms of beneficiary contribution and financial institutions were quite satisfactory.

Table 3.1: Financing pattern of projects of different local bodies in Thiruvananthapuram district 1997-98

(*Rs. in crores*)

Local Body	*No. of Projects*	*Grant-in-Aid*	*Own Fund*	*State Sponsored*	*Centrally Sponsored*	*Loans from Co-operatives*	*Loans from Financial Institute*	*Voluntary Contribution*	*Beneficiary Contribution*	*Others*	*Total Expen-diture*
Gram Panchayat	4125	39.70	3.30	1.48	1.09	0.96	7.01	3.32	25.36	1.11	83.33
%		47.60	3.90	1.78	1.30	1.15	8.42	3.98	30.44	1.33	100.00
Block Panchayat	649	10.05	0.10	0.40	3.60	0.18	1.17	0.54	2.82	0.22	19.09
%		52.60	0.50	2.11	18.8	0.93	6.12	2.83	14.78	1.18	100.00
District Panchayat	225	10.89	0.02	0.00	0.00	0.63	0.60	1.41	1.83	0.08	15.44
%		70.48	0.13	0.00	0.00	4.05	3.89	9.10	11.83	0.52	100.00
Corporation	131	12.67	1.41	5.25	0.92	0.73	8.84	2.62	8.18	0.68	41.30
%		30.60	3.40	12.7	2.22	1.77	21.40	6.34	19.80	1.66	100.00
Municipality	294	3.83	2.61	0.64	0.27	0.06	4.78	0.47	1.53	0.52	15.34
%		26.03	17.73	4.38	1.84	0.43	32,52	3.18	10.38	3.52	100.00

Source: Constructed from Appendix Forms of Local bodies, 1997-98. State Planning Board (SPB)

Table 3.2: Category-wise financing pattern of projects of local bodies in Thiruvananthapuram district 1997-98

(*Rs. in crores*)

Local Body	*No. of Projects*	*Grant-in-Aid*	*Own Fund*	*State Sponsored*	*Centrally Sponsored*	*Loans from Cooperatives*	*Loans from Financial Institute*	*Voluntary Contribution*	*Beneficiary Contribution*	*Others*	*Total Expenditure*
General	3932	53.09	6.93	7.26	5.67	2.12	19.87	7.34	33.63	2.38	138.20
%		38.39	5.01	5.25	4.10	1.53	14.37	5.31	24.32	1.72	
SCP	1333	22.21	0.48	0.52	0.20	0.44	2.38	0.92	5.83	0.24	33.23
%		66.86	1.45	1.56	0.61	1.32	7.17	2.76	17.54	0.73	100.00
TSP	159	1.83	0.02	0.00	0.01	0.00	0.15	0.09	0.26	0.00	2.36
%		77.29	1.05	0.02	0.44	0.00	6.52	3.83	10.85	0.00	100.00
Total	5424	77.13	7.44	7.78	5.88	2.56	22.40	8.35	39.71	2.62	173.87
%		44.36	4.28	4.48	3.38	1.47	12.89	4.80	22.84	1.51	100.00

Source: Constructed from Appendix Forms of Local bodies, 1997-98. State Planning Board (SPB)

The dependence on Grant-in-aid by the local bodies is closely linked to the ability of the local bodies to tap additional resources. Among all the local bodies the lowest share of Grant-in-aid was for municipalities, i.e. 26.03 per cent and highest in the case of District Panchayats i.e. 70.48 per cent. The lowest share of Grant-in-aid in municipalities is due to the possibility for tapping more credit from financing institutions compared to other local bodies. It is further seen from the data that white the Grant-in-aid had the largest ratio i.e., 70.48 per cent, the contribution from credit institutions for district panchayat was only 3.89 per cent.

From the table it is clear that local bodies have to depend on financial institutions wherever the state assistance is low. For example, when Grant-in-aid was 26.03 per cent, loans from financial institutions declined to 21.40 per cent. Further, when Grant-in-aid was 47.64 per cent and 52.65 per cent for Gram panchayat and Block panchayat respectively, the additional tapping of resources from credit institutions has come down to 8.42 per cent 6.12 per cent. Finally, with highest ratio of Grant-in-aid of 70.48 per cent, the need for additional resources in the form of credit was only 3.89 per cent. This shows that loans from financial institutions play an important role where state assistance is less for the successful implementation of the projects.

The category-wise financing pattern of projects shows that out of 5424 projects, 3932 projects are under the general category. Naturally the state assistance in the form of Grant-in-aid and loans from financial institutions are high compared to Special Component Plan (SCP) and Tribal Sub Plan (TSP) categories. Out of total outlay of Rs. 173.87 crores the allocation for general, SCP and TSP categories are Rs. 138.28 crores, Rs. 33.23 crores and Rs. 2.36 crores respectively. The total expected loans from financial institutions was Rs. 22.40 crores. Out of this major amount, Rs. 19.87 crores has gone for general category. The SCP and TSP received only Rs. 2.38 crores and Rs. 0.09 crores respectively. This is because compared to SCP and TSP Grant-in-aid or state assistance is low for general category. This is clear from the table that out of total expenditure for each category the percentage share for General, SCP and TSP categories are 38.39 per cent, 66.86 per cent and 77.29 per cent respectively. Consequently, the expectation of additional resources from financial institutions is very high in the general category.

There are three sectors under which projects are formulated in the decentralized planning. It is interesting to analyze the sector-wise financing pattern of projects of different local bodies (Table 3.3). The analysis of results presented in the table gives us an idea about the allocation of finance among productive, service and infrastructure sectors at all levels of local bodies.

The maximum number of projects (4125) are in gram panchayat. In the gram panchayat state assistance is more in the productive sector, i.e. Rs. 15.48 crores. In the service sector it is Rs. 14.32 crores and in infrastructure it is only Rs. 9.53 crores. It is because the sectors with larger number of projects receive higher state assistance. However, the grant-in-aid as a percentage of total outlay for each sector shows that productive sector got only 33.94 per cent as grant-in-aid out of Rs. 45.60 crores. In the gram panchayat the success of projects depends upon the additional resources available from financial institutions and beneficiary contribution. For the service and infrastructure sectors in the gram panchayat the expectation of credit from financial institutions is not high.

In the block panchayat the expected contribution from financial institutions for productive and infrastructure sectors were 9.27 per cent and 5.96 per cent respectively. However, for infrastructure sector there was no expectation of credit from financial institutions.

In the district panchayat, it is the service sector where the credit is very high. However, the state assistance is very high in district panchayat. The percentage shares of grants-in-aid in the total outlay for productive, service and infrastructural sectors are found to be 61.86 per cent, 79.29 per cent and 61.72 per cent respectively. Thus, the percentage of institutional finance plays an important role. The state assistance in the form of grant-in-aid is very low in municipality and corporation. For example, in municipalities the contributions to these three sectors from their respective total outlay were 21.30 per cent, 52.51 per cent and 6.45 per cent respectively. Thus, the loans from financial institutions play a major role in the successful implementation of these projects. Thus, in the municipality the expected contribution of the financial institutions for productive, service and infrastructure sectors were 42.88 per cent, 5.91 per cent 36.45 per cent respectively.

In the corporation there is no expectation of credit in the infrastructure sector. But for the productive and service sectors the expected contribution was 32.24 per cent 20.66 per cent respectively. The state assistance was not enough to meet the requirements for developmental projects.

Table 3.3: Sector-Wise Financing Pattern of Projects of Different Local Bodies in the Tvpm. district 1997-98

(Rs. in Crores)

Local Body	*Sector*	*No. of Projects*	*Grant-in-aid*	*Own Fund*	*State Sponsored*	*Centrally Sponsored*	*Loans from Co.*	*Loans from Financial Institute*	*Voluntary Contribution*	*Beneficiary Contribution*	*Others*	*Total*
Gram panchayat	Productive	1697	15.48	0.50	0.54	0.08	0.83	6.26	0.71	20.66	0.53	45.60
	%		33.94	1.09	1.18	0.19	1.84	13.73	1.55	45.32	1.17	100
	Service	1299	14.32	1.27	0.55	0.66	0.12	0.74	1.04	4.02	0.42	23.15
	%		61.86	5.49	2.37	2.85	0.53	3.18	4.49	17.38	1.83	100
	Infrastructure	1129	9.53	1.36	0.38	0.32	0.00	0.02	1.21	0.61	0.13	13.56
	%		67.90	10.51	2.72	2.34	0.00	0.11	10.77	4.59	1.05	100
Block panchayat	Productive	232	3.26	0.00	0.04	0.60	0.16	0.58	0.15	1.47	0.03	6.33
	%		51.61	0.00	0.77	9.60	2.65	9.27	2.29	23.30	0.50	100
	Service	285	4.60	0.00	0.34	2.69	0.01	0.58	0.21	1.25	0.04	9.73
	%		47.27	0.00	3.58	27.64	0.10	5.96	2.17	12.80	0.47	100
	Infrastructure	132	2.17	0.09	0.005	0.30	0.00	0.00	0.18	0.09	0.15	3.02
	%		72.19	3.27	0.20	10.11	0.00	0.00	6.09	3.29	4.89	100
District panchayat	Productive	85	3.19	0.02	0.00	0.00	0.00	0.10	0.19	1.57	0.08	5.15
	%		61.86	0.39	0.00	0.00	0.00	1.97	3.63	30.61	1.55	100
	Service	86	6.07	0.00	0.00	0.00	0.62	0.50	0.22	0.23	0.00	7.65
	%		79.29	0.00	0.00	0.00	8.16	6.53	2.93	3.09	0.00	100
	Infrastructure	54	1.61	0.00	0.00	0.00	0.00	0.00	0.99	0.01	0.00	2.62
	%		61.72	0.00	0.00	0.00	0.00	0.00	37.90	0.38	0.00	100

(Contd.)

Table 3.3 (*Cantd.*)

Local Body	*Sector*	*No. of Projects*	*Grant-in-aid*	*Own Fund*	*State Sponsored*	*Centrally Sponsored*	*Loans from Co.*	*Loans from Financial Institute*	*Voluntary Contribution*	*Beneficiary Contribution*	*Others*	*Total*
Municipality	Productive	87	1.27	0.31	0.19	0.00	0.05	2.49	0.04	1.21	0.28	5.83
	%		21.30	5.38	3.40	0.00	0.91	42.88	0.66	20.71	4.76	100
	Service	120	1.63	0.38	0.44	0.05	0.01	0.18	0.11	0.91	0.09	3.12
	%		52.51	12.35	14.27	1.60	0.32	5.91	3.59	6.37	3.08	100
	Infrastructure	87	0.94	190	0.00	0.22	0.00	2.10	0.31	0.12	0.14	5.76
	%		6.45	33.15	0.00	3.82	0.00	36.46	5.52	2.09	2.51	100
Corporation	Productive	35	3.37	0.33	3.03	0.37	0.58	4.91	0.27	2.37	0.01	15.25
	%		22.09	2.19	1.7	2.42	3.80	32.24	1.74	15.56	0.07	100
	Service	60	7.19	0.62	1.09	0.54	0.15	3.91	2.29	2.69	0.42	18.95
	%		37.98	3.32	5.79	2.89	0.80	20.66	12.12	14.21	2.24	100
	Infrastructure	36	2.10	0.45	1.12	0.00	0.00	0.00	0.06	3.11	0.25	7.08
	%		29.65	6.33	15.81	0.00	0.00	0.00	0.78	43.91	3.52	100

Source: Construction from Appendix Forms of Local bodies, 1997-98. State Planning Board (SPB)

Thus, from the Table 3.3 it can be seen that except for district panchayat the need for additional local resources in the form of loans from financial institutions is necessary for the successful implementation of the development projects.

Table 3.4 gives the details of credit expectations for sub-sectors like agriculture, industry, social welfare etc. It tells us that the sectors like agriculture, soil conservation, industries, housing, social welfare, infrastructure etc. have very high credit expectations. The expectation of credit is the highest in agriculture i.e. Rs. 7.55 crores followed by that of industry and soil conservation which were Rs. 3.24 crores and Rs. 3.59 crores respectively. Education, Health, General Services, Energy, Transport etc. are the sectors in which loans from financial institutions is low. This may be because of large percentage share of state assistance in the form of grant-in-aid in their respective total outlay.

The total number of projects in the year 1997-98 was 5424 with a total outlay of 173.87 crores. Out of this 372 projects are having credit linkages or expected contribution from financial institutions with a total outlay of Rs. 44.75 crores.

In the table 3.5 the financing pattern of projects which are having credit linkages or expected contribution from financial institutions for every local body is given. It is evident from the data that out of Rs. 44.75 crores almost half of the expected contribution was by loans from financial institutions. The grant-in-aid was only Rs. 9.31 crores. Beneficiary contribution was also as low as Rs. 6.16 crores. It shows the importance of the role of credit for the successful implementation of the projects. In all local bodies pattern was the same.

Realized or Final Expenditure Pattern of Projects

The expected amount of financing in the year was Rs. 173.87 crores. But the realized or final expenditure is only Rs. 56.64 crores. The category-wise expenditure pattern shows that there is a wide disparity between expected and realized amount. The expected amount for General, SCP and TSP categories were Rs. 138.28 crores, Rs. 33.23 crores and Rs. 2.30 crores respectively. However, the realized amounts are only Rs. 43.22 crores, Rs. 12.99 crores and Rs. 0.53 crores, which show a gap of Rs. 95.06 crores, Rs. 20.29 crores and Rs. 1.83 crores respectively. The realized amount from financial institutions was Rs. 0.53 crores in the case of General category

Table 3.4: Expected Financing Pattern of Projects for the Year 1997-98 (Sector-wise Allocation)

(Rs. in crores)

Sector	*Grant-in-Aid*	*Own Fund*	*State Sponsored*	*Centrally Sponsored*	*Loans from Co-operatives*	*Loans from Financial Institutions*	*Voluntary Contribution*	*Beneficiary Contribution*	*Others*	*Total Outlay*
Agriculture	17.36	0.50	0.71	0.34	1.01	7.55	0.52	24.93	0.84	53.76
Soil conservation	4.50	0.22	2.60	0.52	0.04	3.24	0.66	1.18	0.01	12.97
Industries	4.69	0.45	0.51	0.20	0.58	3.59	0.16	1.18	0.08	11.44
Education	5.40	0.57	0.61	0.22	0.001	0.002	2.24	1.41	0.25	10.70
Health	11.48	0.86	0.73	0.75	0.01	1.57	0.84	2.65	0.48	19.37
Housing	12.19	0.41	0.34	0.47	0.01	2.18	0.30	3.49	0.15	19.54
Social welfare	4.15	0.27	0.42	2.33	0.35	2.07	0.44	0.76	0.06	11.20
Gen Services	0.66	0.18	0.34	0.16	0.55	0.01	0.07	0.10	0.04	2.11
Energy	2.71	0.57	0.13	0.18	0.00	0.01	0.11	1.18	0.35	5.25
Transport	13.48	2.21	1.39	0.43	0.00	0.03	2.99	2.80	0.33	23.63

Source: Constructed from Appendix Forms of Local bodies, 1997-98. State Planning Board (SPB)

Table 3.5: Local Body-wise Financing Pattern of Projects Which have Credit Linkages or Expected Contribution from Financial Institutions in the Thiruvananthapuram District 1997-98

(*Rs. in crores*)

Local body	*Grant-in-Aid*	*Own Fund*	*State Sponsored*	*Centrally Sponsored*	*Loans from Co-operatives*	*Loans from Financial Institutions*	*Voluntary Contribution*	*Beneficiary Contribution*	*Others*	*Total Outlay*
Gram panchayat	3.60	0.12	0.23	0.166	0.11	7.01	0.13	3.86	0.006	15.23
Block Panchayat	0.44	0.00	0.05	0.06	0.02	1.16	0.00	0.35	0.006	2.13
District	0.18	0.00	0.00	0.00	0.00	0.60	0.00	0.10	0.00	0.89
Corporation	4.47	0.10	3.13	0.13	0.05	8.83	0.77	1.08	0.00	19.05
Municipality	0.62	0.80	0.15	0.10	0.004	4.47	0.82	0.77	0.03	6.96
Total	9.31	1.02	3.56	0.45	0.634	22.07	0.92	6.16	0.42	44.75

Source: Constructed from Appendix Forms of Local bodies, 1997-98. State Planning Board (SPB)

and Rs. 0.39 crores for SCP. In the case of TSP there was no loan from financial institutions. Compared to the expected loans for Rs. 19.87 crores, Rs. 2.38 crores and Rs. 0.15 crores for these three sectors, the realized amount is far from satisfactory. This is shown in Table 3.6.

Table 3.6: Category-wise Expected and Realized (final) Credit from Financial Institutions for the Projects in the Year 1997-98

Category	*Loans from Financial Institutions*		*Total Outlay*	
	Expected	*Realized*	*Expected*	*Realized*
General	19.87	0.53	138.28	43.27
SCP	2.38	0.39	33.23	12.99
TSP	0.15	0.00	2.30	0.53

Source: Constructed from Appendix Forms of Local bodies, 1997-98. Kerala State Planning Board (SPB)

The spatial distribution of loans from financial institutions shows wide difference between different blocks. Out of 12 blocks, 5 blocks did not get any loans from financial institutions. The maximum credit was obtained by 3 blocks. Out of total contribution from financial institutions, i.e. Rs. 0.64 crores, Varkala, Nemom and Perumkadavila alone had got Rs. 0.50 crores. The remaining amount of Rs. 0.14 crores was received by another 4 blocks. This shows a wide disparity between spatial distribution of loans from financial institution. This is shown in Table 3.7.

In Thiruvananthapuram district, out of 5424 projects 372 projects expected credit from financial institutions. But only 29 projects could get finance from financial institutions. The details of the expected and realized amount of credit for these 29 projects are given in Table 3.8

For the above 29 projects the percentage of realized credit was 100 per cent. The projects from all the sectors got credit from financial institutions. Out of these 29 projects, for three projects expected contribution from financial institutions was more than 70 per cent. Seven projects expected credit for more than 50 per cent. The number of projects whose expectation of credit is less than 25 per cent is only two. This shows that assistance from financial institutions is an important factor.

Table 3.7: Block-wise final expenditure pattern of projects in Thirvananthapuram District

Block	*Grant-in-Aid*	*Own Fund*	*State Sponsored*	*Central Sponsored*	*Loans from co-operatives*	*Loans from Financial Institutions*	*Voluntary Contribution*	*Beneficiary Contribution*	*Others*	*Total*
Varkala	2.65	0.00	0.02	0.00	0.00	0.21	0.10	1.08	0.00	4.06
Kilimanoor	3.34	0.00	0.00	0.00	0.00	0.00	0.86	0.53	0.008	4.73
Chirayinkil	3.11	0.04	0.002	0.009	0.00	0.00	0.13	0.56	0.04	3.89
Vamanapuram	3.56	0.05	0.004	0.002	0.00	0.00	0.004	2.14	0.05	5.81
Vellanadu	3.43	0.07	0.00	0.00	0.00	0.00	0.09	1.64	0.007	5.23
Nedumangadu	1.92	0.001	0.00	0.03	0.002	0.01	0.28	1.44	0.00006	3.58
Kazhakuttam	4.09	0.04	0.001	0.00	0.003	0.05	0.46	1.22	0.08	5.94
Tirvandrum rural	2.40	0.04	0.01	0.02	0.05	0.03	0.44	0.40	0.002	3.39
Nemom	4.19	0.23	0.14	0.03	0.00	0.13	0.60	1.02	0.009	6.35
Perumkadavila	3.28	0.18	0.05	0.06	0.00	0.16	0.11	1.42	0.01	5.27
Athiyannoor	2.70	0.02	0.01	0.00	0.00	0.00	0.02	1.04	0.00	3.82
Paraggala	3.15	0.05	0.01	0.03	0.00	0.04	0.26	0.93	0.00	4.47
Total	37.79	0.73	0.25	0.18	0.08	0.63	3.41	13.39	0.21	56.64

Source: Constructed from Appendix Forms of Local bodies, 1997-98. State Planning Board (SPB)

Table 3.8: Expected and realized amount from financial institutions—Panchayat-wise (29 projects)

Panchayat/Project Name	*Expected Amount*	*Realized Amount*	*Total Amount*	*% of Realized*	*% of Expected to Total*
Chemmaruthy					
One Cow for one house prog.	176000	176000	650034	100	27.07
Well Irrigation	15000	15000	225000	100	6.66
Cherunniyoor					
Milk producing unit	18000	18000	64000	100	28.12
Milk producing unit	27000	27000	96000	100	28.12
Poultry Farming	66000	66000	93000	100	70.96
Employment creation prog.	100000	100000	350000	100	28.57
One Cow for one house	194000	194000	213600	100	90.82
Integrated paddy development Programme	28000	28000	687422	100	40.73
Pothencode					
Panchayat level chlorination prog.	25000	25000	99000	100	25.25
Irrigation for Vegetable cultivation	125000	125000	250000	100	50.00
Irrigation for coconut cultivation	125000	125000	250000	100	50.00
Sreekariam	24200	24200	48080	100	50.33
Vellarada					
Cow Farming	25500	25500	131050	100	19.46
Egger nursery	29165	29165	47165	100	61.83
Development of paddy cultivation	100000	100000	897300	100	11.14
Cultivation of medicinal plants	100000	100000	300000	100	33.33
Cow Farming	120000	120000	364000	100	61.54
Vegetable cultivation	511403	511403	2049950	100	24.95
Vilavoorkkal					
Vegetable cultivation	62500	.62500	148500	100	42.09
Mini dairy unit	90000	90000	272000	100	33.09
Poultry Farming	289500	289500	529485	100	54.67
Goat Farming	120000	120000	195000	100	61.54
Malayankeezhu					
Coconut cultivation	11225	11225	311600	100	36.02
Distribution of Sewing machines	120000	120000	195000	100	57.14
One Cow for one house	240000	240000	364000	100	65.93
Thirupuram					
Production of ready made garments	200000	200000	351000	100	56.98
Nemom					
Orchid cultivation	225000	225000	742000	100	30.32
Mangalapuram					
Cow Farming	240000	240000	616150	100	38.95

The foregoing analysis shows that the credit from financial institutions was very low for the implementation of projects under decentralized planning. The realized amount lags behind the expected credit for majority of the projects. This hampers the implementation of development plans under decentralized planning. The credit planning authority should take steps to integrate the credit planning with the development planning under decentralized planning in Kerala. For this, Village Credit Plan (VCPs) should be formulated under the Service Area Approaches (SAPs) of individual banks in Kerala.

Preparation of credit plans implicitly involves preparation of development plans. A credit plans presupposes the existence of a development plan so that the former can subserve the objectives of the latter. Unless there is a complete collaboration and co-ordination between the two, there might arise a confusion of roles between bankers and Government functionaries. As a result, development plans would effectively become conditional upon credit plans, rather than the other way round. Preparing credit plans in isolation and without proper co-ordination with the development plans could be counter productive. It may be necessary, therefore, to specify in more detail their respective roles and collaborative arrangements for preparing the plans more specifically so as to avoid over-lapping at the field level. What is more important is that bankers and development agencies prepare their respective plans jointly. Such collaborative aspect of rural planning should be given added emphasis under the new approach. Credit planning is ineffective if it is not backward linked with a development plan in physical terms and if it is not forward linked with, via credit disbursing agents—the farmers, the village professionals and artisans, the rural industrialists and the infrastructure builders, both governmental and private (RBI 1989). As the existing micro level credit planning has been showing signs of slackness with the decentralized planning, despite several institutional innovations and strengthening measures initiated in the past, a relook is essential at our rural credit planning system as to whether the existing system can cope with the latest developments including decentralized planning.

While decentralized planning through the allocation of real resources aims at a rational and proper allocation of factors of production to maximize output, income and employment, the essence of financial planning is to ensure that the demands and supplies are matched in a manner which exploits major and unplanned changes in the price structure. In other words, there has to be a proper integration between physical and financial resources at the sectoral, regional and micro levels which is conducive to achieving contemplated rate, pattern and composition of growth and other distributional objectives of the decentralized planning. Credit planning is, therefore, essentially complementary to decentralized planning so that there is, a balance between financial demands for various kinds of goods and services and their supplies, savings and investment. As institutional credit accounts for a significant proportion of the financial resources available in the economy, the credit planning should be in conformity with the overall decentralized plan in financial and physical terms.

Decentralized decision making has a potential for reducing costs and increasing efficiency. Efforts to enhance decentralization of business procedures, in particular the authority to approve loans, however, are often resisted in financial institutions because of a concern about preventing misappropriation of funds. In general, decentralization may be resisted because of bureaucratic precedent that favours hierarchy, centralization of power and restricted flows of information. However, decentralization should be explored as an alternative that can improve performance, save time, empower staff, create profit and perhaps most importantly, increase the reach of institutions of lower – income households.

Decentralization is also increasingly required in a modern or modernizing economy in which competition is a driving force for change and efficiency. Decentralization requires transparency and accountability as well as sanctions against insiders who abuse it.

Credit decision making can be decentralized if branch managers and lending officers are given clear guidelines and procedures, if there are effective controls and individual lending limits, and if management information systems are sufficiently robust to permit measurement of performance and results. The advantages of decentralized credit decisions include lower transaction costs for all concerned and loans that are sanctioned on the basis of information available to branch decision-makers. The decentralization of credit planning is not a problem in a state like Kerala, where the decentralization of development planning already exists. Like decentralized planning from below, the general assembly or meetings of the entire membership in credit institutions can be used to obtain guidance about the nature of demand for financial services, and acceptable terms and conditions. This gives member of low income households or rural poor an opportunity to participate in decision making and express their concerns and needs to others.

In a credit system, which is becoming increasingly sophisticated, its present weaknesses notwithstanding, and is attempting to bring qualitative improvements in its lending operations, credit planning assumes crucial importance. Such credit planning has to be undertaken not merely at the apex level or even intermediate levels like the regional offices of the credit institutions but more importantly at the grass root level as well, viz. bank branches and co-operative institutions. But credit planning and the blue prints prepared for such planning from time to time seem to have so far made no serious dent at the level of actual formulation and implementation of credit plans (RBI 1988/9). However, in the present decentralized planning system, the quality of statistics required for credit planning at the grass root level is satisfactory. Earlier little data were available about production, marketing, equipment, income-estimates, transportation, village and cottage industries-most of the sectors for which the credit plans are to be formulated. Thus, now it is the high time to co-ordinate rural credit agencies, banks and development agencies to formulate the village credit plans in order to integrate the financial planning and the development planning under decentralized planning in Kerala.

NOTES AND REFERENCES

1. In the present study financial planning means credit planning of banks under the District credit planning system.
2. Kotaiah, D.P., (1997): Banking on micro entrepreneurs—Innovative lending and credit programmes, *National Bank News Review*. April-May 1997.
3. Potential Linked Credit Plan is a comprehensive documentation of potentials for rural economic activities, both in physical and financial terms. It is an assessment of the gaps in infrastructural support which need to be filled in, in order to fully exploit the realisable potentials.
4. Isaac, Thomas, (1999): *People's Planning: Towards a Hand Book,* State Planning Board, Thiruvananthapuram.
5. Reserve Bank of India (1989): "Report of the Agricultural Credit Review Committee", p. 419.
6. *Ibid,* p.

4

People's Plan for Sustainable Rural Development

SITAKANTA SETHY

Introduction

On 15th August, 1947 at the midnight hour when the world slept, India awoke to "life and freedom". But after fifty years of its independence when some developing countries have achieved the status of "East Asian Tigers", rural India is still sleeping. Starvation, deaths, illiteracy, high infant mortality rate are still pervasive and prevalent. Self-regeneration, self-help and self-reliance are still miracles for rural people. A co-ordinated effort from the state, the legislature, the political executive and bureaucracy is expected to tap the potentiality of rural areas and the poor. Destitution hurts, degrades and drives people into desperation. It is a wrong notion with the planners and economists that rural people being illiterate may not be able to plan. But by NDGOs intervention rural people have proved that in local conditions, with local resources and with grassroots leadership they can be planners as well as better managers.

Poverty is both an absolute and a relative concept. Rural people who are below poverty line without having basic necessities in life to meet their ends come under absolute poverty. Sometimes people consider it as God's gift or the unjust construct of society. People in India are materially poor but ethically rich. Poverty is undesirable, so people who are poor have to plan for themselves and their area. An integrated approach by development administration, non-government developmental organizations (NGDOs), the rural poor and the Panchayati Raj Institutions is necessary to formulate "people's plan" for sustainable rural development.[1]

Rural Development Planning and the Poor

More than seventy per cent of people in India live in rural areas and agriculture is the main source of their income and the percentage of the rural people lying below the poverty line is around 50. For rural development, planning and administration there is a Ministry for Rural Areas and Employment in which there are special departments like Department of Rural Development and Department of Rural and Poverty Alleviation. During the first five year plan emphasis was on community development, in the second it was on co-operative farming and in the third plan Panchayati Raj model was adopted while in the subsequent plans, i.e., the 4th, 5th, 6th and the 7th plans, emphasis was given on area-based development, minimum needs programme, anti-poverty, wage and self-employment. After the 7th plan it was found that rural development was not picking up. Development benefits were neither reaching the people nor the "trickle down effect" was realized. In the eighth five year plan special emphasis was given on renovation of Panchayati Raj Institutions as a measure of decentralization of power.[2]

Till the Constitution 73rd Amendment, planning from above or "top-down" approach was operating. Poverty alleviation programmes claim to be based on assessment of 'peoples' need". India has progressed gradually in designing such programmes. It has been a process of "learning by doing" in the absence of a structured model of poverty alleviation suited to the heterogeneous character of poverty which exists in India. Planning for rural development has been on the basis of rural needs and priorities as perceived by the urban planners and their "world view" of development has been administered at grass root level. Their priorities of the rural poor, however, failed to give satisfactory results in this direction.[3]

Characteristics of the Rural Poor

Rural poor in India have peculiar characteristics which mismatch the common belief of urban planners and economists. Rural people are mostly dependent on agriculture. Being landless they have to work as agricultural labourers whose earnings are very low and uncertain. Having big families and low income they suffer from ill-health and indebtedness. Due to malnourishment infant mortality rate (IMR) is very high. Being illiterate they are unable to know their outside world, which results in their weak bargaining capacity. Characteristics like proneness to dependency, immediate need gratification, fatalistic attitude and naive and magical consciousness are also found with the rural poor and these do not help them to come out of the "vicious circle of poverty". The above characteristics are liabilities for them (see Table 4.1.)

Table 4.1: Liabilities

• landlessness	• low & uncertain income
• big family	• indebtedness
• immediate need gratification	• dependence-proneness
• low life expectancy	• weak bargaining capacity
• illiterate	• isolated household
• malnourished children	• fatalistic attitude
• naive and magical consciousness	• caste & class division
	• ill health

Inspire of the above liabilities, rural people possess some characteristics (see Table 4.2) which are assets for them which may be utilized for better planning and implementation of different poverty alleviation programmes by the government agencies. Rural people believe in community life and very often use traditional medicines. Their friendly attitude towards man and nature, particularly cultural practices, help them to live in peaceful environment. Spiritual beliefs and God-fearing nature discourage them to pass biased and unfair judgments at village panchayats. In group discussions and village meetings they take right decisions for the development of their area and people, outside influence hardly influences them. They believe in hard labour and simple living. Until and unless they come across unsustainable urban life, they do not think that they are poor. Some times government programmes are dumped upon them which is illogical and unethical. In the name of development all-weather roads are made. The motive behind it is to cut down forest and make their life more miserable and unsustainable.

Table 4.2: Assets

• community life	• cultural practices
• collective decision making	• simple and happy life
• common property resources	• social control
• believe in traditions	• brotherhood
• hospitality	• justice
• hard labour	• spiritual beliefs

Some people-oriented non-governmental developmental organizations (PINGOs) identify the potentials, i.e., liabilities and assets of rural poor of certain areas, sit, discuss and plan with them for sustainable rural development. In the process their facilitators facilitate to convert their liabilities into assets; which is really a challenging task.

Development Administration and the Rural Poor

Rural Development is an interminable process. Even the partial success of a set of programmes takes rural society forward. Generally speaking, rural development administration has a low perception of its tasks and it is plainly more job-oriented than service-minded. Its most impressive achievement is that it keeps its records well to evade any adverse action. These have over the years become effective criteria of success. The bureaucracy has no mechanism of self-correction and the successive plans and approaches to rural development disclose a verily similar pattern. The need is to find an alternative to bureaucracy for implementing these programmes.[4]

Rural poor living far away from the development administration could neither reach the authorities for assistance nor the authorities reached them. So a big gap is maintained between the two, which helps to promote "mai-bap" image. Whenever people come for assistance, they are harassed in many ways. People from development administration do not visit their area because of lack of communication and motivation. They develop a tendency of self-justification. In order to achieve target in time they very often try to impose programmes during the end of the financial year in a hurry and the unspent money is returned. Skill and experience of district level officers do not get a chance to be utilized. Lack of political and bureaucratic "will" to implement programmes properly help to do development works on pen and paper. Departmental accountability does not encourage proper implementation of various developmental programmes. Monitoring and evaluation reports of the programmes implemented are hardly published.

People for whom plans and programmes are meant, by whom it should have been implemented are no where in the process. If development is "development of people and area which enhances their ability to make right decisions at right time and to shape their own future in the light of their enlightened understanding of their environment" then are the rural poor really participating in development? Is the rural development administration able to involve them in the development process? However, the wide spread impression is that benefits of the developmental programmes have not accrued to the rural poor to the desired extent. Many deficiencies have surfaced in the implementation process. The so-called 'delivery system' has never thought of strengthening their own system so that people at the other end will be able to utilize the resources. This feeling has been increasingly strengthened over the years, i.e.,

- people are "objects" rather than "subjects" to change;
- trend towards "evidences" in implementation of developmental programmes;
- emphasis on target achievement rather than substantive results;
- lack of concern for the poor with the bureaucracy;
- lack of pubiication of monitoring and evaluation reports;
- need of social accountability in development programme implementation;
- denial of access over public information of Official Secrets Act.

Moreover, rural development administration has some weaknesses which could be converted into strengths for a transparent and reliable system. Development administration never takes moral responsibility for failure of any programme; rather a self-justifying behaviour is gaining momentum day by day (see Case 1). Decision to implement a programme in a certain area, its budget and the middlemen who will implement are very often influenced by narrow politics as a result of which village communities never get any programme at all. The Official Secrets Act has encouraged corruption by not giving people a chance to see the details of development expenditure. Target achieving mentality, bureaucratic control, lack of innovation and, above all, lack of motivation to work do not encourage successful implementation of programmes for poverty alleviation, promotion of self-help and self-reliance.

SWOT Analysis—Case 1: Development Administration

Strengths

- central and state resources
- constitutional legitimacy

- downward power strategy
- bureaucratic control
- accountable to authority
- trained and untrained manpower

Weaknesses

- "mai-bap" image
- palm greasing
- political pressure on decision
- piecemeal approach
- year-end programme implementation
- mass mobilization
- resource tapping
- partial utilization of assistance

Opportunity

- time and target achievement
- achievement motivation
- achieve moral and departmental responsibility
- power play and job satisfaction
- improve technical competency
- sustainable development

Threat

- self-justification
- programme implementation in pen and paper
- middlemen in implementing programmes
- criminal, politician and bureaucracy nexus

Development administration has enormous financial resources, trained and experienced manpower. By proper use of downward power strategy rural development administration can plan and implement "need based" programmes successfully. Self-justification for wrong choice of beneficiary, showing targets on pen and paper, ignoring rural poor and encouraging middlemen to implement programmes and, above all, nexus between criminals and politicians in the implementation of development programmes are growing day by day which need special attention and care. Mass campaign by NGDOs and development administration to involve people at village level to plan, implement and evaluate may solve the problem.

Panchayati Raj and the Rural Poor

Central Planning failed to give rural development a momentum. So again attempt was made to organize and regenerate Panchayati Raj Institutions. The concept behind it was that through decentralization of power, planning and implementation of developmental programmes will be done at the grassroot level which will encourage people's participation and empower women for micro-level planning.[5] But people having no education, experience in planning, divided between *caste and class* seem to be helpless in planning and implementing programmes (see Case 2). PRIs depend on both state and centre for funding, they are still unable to collect taxes and manage their affairs themselves. The people who are supposed to run the office are untrained, illiterate and lack critical understanding about different development programmes, which are also being implemented by them. Block office is the only resource centre for them.

SWOT Analysis—Case 2: Panchayati Raj Institutions

Strengths

- access to resources
- charismatic leadership
- small area of operation
- constitutional legitimacy
- political participation

Weakness

- untrained manpower
- politics as divisive force
- caste-class division
- gender bias
- weak decision-making power
- experience in planning

Opportunity

- public hearing to check corruption
- gram sabha as sub-regional planning body
- pressure groups
- people's organization
- no dependence on external finance
- local initiative
- learning by doing

Threat

- no reliable data base for planning
- discrimination by elected leaders
- self-justification tendency of V.L.W. (village level workers)
- dissolution power with legislature

Very often inter-and intra-party politics, class and caste division influence the decisions of the PRIs as a result of which the poorest of the poor suffer much. Since long rural elite were ruling but due to reservation policy, poor and lower caste people got the opportunity to be elected. Having no education, awareness and technical expertise, they have become easy prey of the rural elite and political parties. Above all, without data base, planning by the Panchayati Raj Institutions has become a difficult task, particularly at the Gram Panchayat level. Block Development offices are supposed to play a crucial role in planning, implementation and evaluation. But with so many programmes and with less untrained staff having no motivation for proper documentation block offices have become a mess.

The Panchayati Raj Institutions have access to enormous human and natural resources. With both charismatic and accepted leadership and small area of operation they can plan better for rapid rural development. With regular sitting of gram sabha they can play the role of pressure groups and mobilize both internal as well as external resources for sustainable rural development. Mass organizational base and constitutional legitimacy could be an advantage to run local self-governments effectively. So it is very important to make the Panchayati Raj Institutions more active in planning their own programmes for development. To work effectively and efficiently, they need the co-operation and co-ordination of both the development administration and local NGDOs. For direct and active involvement of the rural poor in development process, planning could be done at village level.

NGDOs and the Rural Poor

The last few decades have witnessed the rising profile of NGDOs – the so called third sector, in development process and policy initiatives. Their participatory and people-centred approaches have scored an edge over other development agencies. It is time to relinquish the idea of NGOs as competitors to public administration and focus attention on their capabilities as innovators in strengthening the service responsiveness, policy making and implementation.[6]

Non-government developmental organizations never stay away from people rather stay with people, try to understand their problem, discuss with them and try to find out the ways to have an easy and permanent solution. That is why the rural people prefer NGDOs to the people from government agencies for immediate help. Good rapport with the community helps them to organize the poor, consciencitize them, promote their leadership and organize them to demand their dues. NGDOs are accountable to people for whom and with whom they work. So there is the least chance for unhealthy practices. Even if it happens, there is a fair chance that people may reject them, since their very sustainability is threatened. People-centred NGOs (PINGOs) strictly stick to their approach to bring a change in people, in capacity building of these people for socio-economic, political and cultural development. NGDO facilitators build awareness among the people to help them, to make them critically conscious about their potentiality and identify the hurdles before them for effecting change. Through spiritual orientation NGDOs try to arouse the feeling component of rural poor to use their brain for changing their socio-economic condition (see Fig. 4.1).

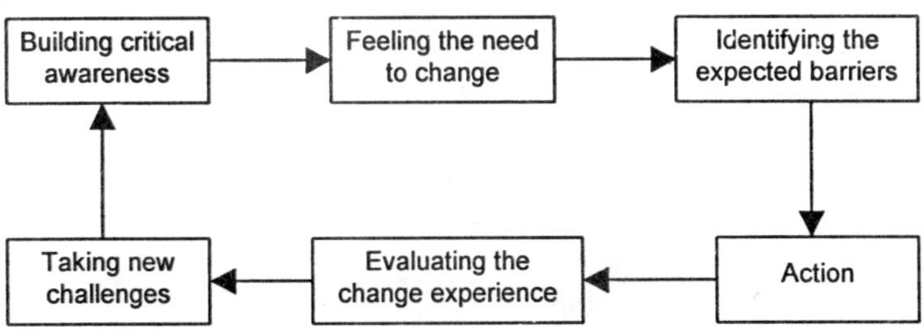

Fig. 4.1: Cycle of Change

People's Organization–I

Broadly, NGOs could be divided into four categories, i.e., Big NGOs, who are less dependent on funds (BINGOs), people oriented NGOs who work with people for their overall development (PINGOs), and some who do research, consultancy, training, networking and funding etc. (RNGOs). For participatory planning and development, co-operation and co-ordination of all the four categories are necessary. Taking strength, weakness, opportunity and threat (SWOT analysis) into consideration NGOs of all the categories needs co-operation from the development administration for their smooth functioning and sustainability (see Case 3). It is a challenging task for NGDOs to extend their area of operation all over the country, bring transparency, networking among themselves and, above all, help the rural poor to plan at village level. Instead of working as separate and independent bodies of "agents of change" they need to work as "agencies of development". Their involvement with PRIs to plan at micro-level may help them to be recognized truly as agencies of development. Since long NGDOs in India are trying to sensitize, organize and mobilize people at village community level and are instrumental in enabling the poor to pursue self-help and self-reliance. A healthy GO-NGO relationship built on mutual trust and co-operation is needed to give rural development a real pick-up.

SWOT Analysis—Case 3: Non-Government Development Organization

Strength

- organization of small and marginal groups
- grassroots leadership promotion
- formation of pressure groups
- popular legitimacy
- dedication and efficiency
- job satisfaction
- result-oriented approach
- missionary zeal
- accepted leadership
- emphasis on ability engineering
- reducing unemployment

Weakness

- overlapping of programmes
- too much paper work
- inner autocracy
- least co-operation from government agencies
- fund dependency
- inconsistency approach
- job security
- transparency in fund utilization

Opportunity

- large scale PRA people's plan
- big networking
- co-operation with government agency
- mass publicity
- class of barefoot managers
- incentives and benefits
- achievement motivation
- diminishing intervention
- innovative work
- appropriate technology promotion
- local resource management

Threat

- following bureaucratic approach
- mushroom growth
- social accountability with unjust people
- more political intervention
- profit motive

Decentralized Planning needs and Integrated Approach

Taking the results of SWOT analysis of development administration, Panchayati Raj Institutions and NGDOs into consideration it may be said that no agency is complete in and by itself. So for sustainable rural development and for giving development a real pick-up planning at the grassroots level, i.e., at the village level, is necessary.

Micro-planning by PRIs will be successful if it could involve people to plan at village level rather that at Panchayat, block and district levels.

Since the First Five Year Plan both development administration and NGDOs are working as separate agencies, as a result of which some patches of development have been possible. Development administration having no access to rural areas people should take the opportunity to work with NGDOs, which very often face the scarcity of funds. For funds NGDOs have to change their approach and area of operation which is never accepted as a good habit. The 1970s saw a spurt in the number of agencies involved in adult education and in the 1980s suddenly they shifted their attention to women's development. So poverty alleviation programmes need an integrated approach which could involve the development administration, NGDOs and Panchayati Raj Institutions and people who need development (see Figure 4.2). Co-operation and co-ordination among the above three agencies is necessary for people's participatory planning and management.

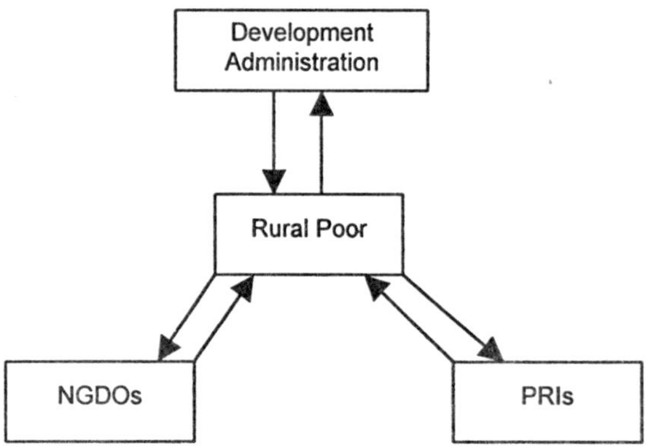

Fig. 4.2: Integrated Approach to Rural Development Planning

People Organization–II

The Seventh Plan paper for the first time envisaged a greater role of the NGOs and voluntary agencies in the implementation of development programmes. These are expected to:

- supplement the efforts so as to offer the rural poor choices and alternatives;
- set an example with limited resources with greater community participation;
- activate the delivery system and to make it effective at the village level to respond to the felt needs of the poorest of the poor;
- disseminate information;
- make communities as self-reliant as possible;
- show how village and indigenous resources could be used;
- find out the ways how the grossly under-utilized human resources, rural technology and knowledge could be used for their development;
- train a cadre of grassroot workers who believe in professionalizing voluntarism;
- mobilize financial resources from within the community and stand on their own feet;
- mobilize and organize the poor and generate awareness to demand quality services and improve a community system of accountability as the performance of village level government functionaries.

The plan paper has put emphasis on NGO intervention in rural development primarily on organizing the poor to demand quality services and impose a community system of accountability on the performance of the village level government functionaries. However, NGDOs are in no way related to lower level planning of the Panchayati Raj Institutions. After the revival and organization of Panchayati Raj Institutions their (the NGDOs') responsibility has increased. Now they can help to:

(*i*) formulate policies and legislations affecting NGDOs themselves, or common people, especially disadvantaged sectors;

(*ii*) carry operational collaboration with NGOs with or without government funding;

(*iii*) develop mutual perception and critique on each other's development initiatives; and (*iv*) carry out peaceful protest movements of people in which NGOs are involved.

The interface between the two is a vertically multi-level and horizontally a multipoint complex covering wide area. In order to make democracy by participation, not by representation, more effective and meaningful and to make development more participative, all NGOs having less financial resources but dedicated, motivated and barefoot managers have an important role to play to make PRIs functional in real terms.

People's Participatory Plan

Our exotic and sophisticated models are professionally convenient in suggesting magic formulae for development and generalized programmes for implementation. But without people's participation plans fail at its inception. NGDOs are known for their virtues of human touch, dedication, flexibility, promotion of self-reliance and nearness to the community. Question may arise as regards how long both the agencies, namely GO and NGO will be working to make rural poor self-reliant; how many families have been lifted above the poverty line so that they do not need help any more. Is not it also important to know whether both the agencies (GO and NGO) are ready for diminishing intervention? Diminishing intervention is a must to make a community independent and this is practised by only a few NGDOs.

Investment of so much of money without any return or a little return is not a welcome idea. For development, more particularly for rural poverty alleviation, funding from external sources could not be considered as a sustainable source for investment. Too many programmes for a pretty long time may help rural poor to develop dependence-proneness and after some years we would need another programme to remove it. Panchayati Raj Institutions, particularly the villages, are the right place to prepare a participatory plan by using participatory rural appraisal (PRA) approach involving people. Development could be sustainable when people will actively participate in the planning, implementation, evaluation and follow-up process to take care of their own environment, problems and prospects (see Fig. 4.3).

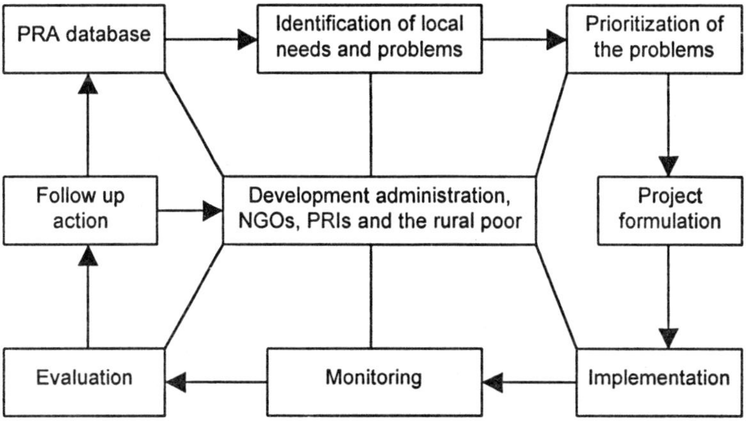

Fig. 4.3: People's Plan for Sustainable Rural Development

People's Organization–III

In a micro-plan Panchayati Raj Institutions, i.e., the elected leaders with the block administration, are supporsed to plan for people, but still rural poor do not have any directed role to play. So, it is more important to involve people in the planning and development process directly, so that they can have the opportunity to plan according to their need. To prepare a "people's plan", preparation of a data base at village level is necessary. Participatory Rural Appraisal (PRA) approach is the most suitable approach to collect and present data by rural people at their respective villages. After collecting and preparing a data base the problems of the people and their area could be prioritised. And according to their priority they can plan and prepare budgets which may include their contribution to solve the problem. Now they can decide how, when and where to implement the projects, who to monitor and evaluate, where and when to discuss the monitoring and evaluation reports and above all, how to follow up the project for better result. Active involvement of development administration, rural poor, PRIs and local NGDOs are necessary to make the process easier and cost effective.

PRA and People's Plan

The origin of PRA as a methodology can be traced back to the multitude of methods and approaches which gradually emerged since the 1970s mainly due to unsatisfactory results obtained from application of conventional methodology to rural development. PRA as a methodology has gained in strength through its emphasis on "participation" in line with paradigmatic shift which is slowly taking place and has helped to fill the vacuum existing in the sphere of rural development. This has also made for its quick spread[8]. PRA is a means of collecting different kinds of data, identifying and mobilizing rural people and evoking their participation and also opening ways in which rural people can participate in decision making, project design, execution and monitoring. At the same time PRA constitutes a process of involvement for indigenous knowledge-building exercises and helps to make timely decision regarding development projects. The participation of rural people in planning, implementing and monitoring of rural development programmes can be facilitated through PRA.

"Planning is the thinking of a group of people poised for development and synchronizing their thoughts with their needs and abilities for anticipated action which will result in their welfare". The above definition of planning reflects that if people at the micro-level have to plan, they should be prepared to think what kind of change they expect and how to materialize it in a more systematic, collective and organized way so that their needs would be fulfilled within stipulated time and fund. Development planning with rural poor at village level is certainly a challenging task. An integrated approach is needed so that NGDOs being close to people, development administration with its resources and PRIs with its infrastructure could be able to bring out a people's plan involving people at the village level.

PRA techniques are still developing to meet the needs of the researchers and planners, but it is the most appropriate language or technique which could be used for participation of people in the planning process at grassroots level. Specialists from development administration, village level workers and elected leaders of PRIs of the area, local people and NGDO facilitators could sit and discuss together to appraise different problems and available resources to solve them at their level. The language which both the villagers and the information seekers can understand and appreciate could be used. Now an illiterate villager can communicate, i.e., pass information, present them through diagrams, symbols or designs. In this way the needs and aspirations of the concerned people could be discussed and their problems could be prioritised. A special art is required to make people participate in the discussion. Without building rapport it is impossible to either collect information or discuss certain topics. So NGDO facilitators' presence cannot be ignored. After information is collected and presented in diagrams, maps or in matrix form, they are again kept for group discussion and analysis in village meeting. Here necessary corrections are made. The most preferred or the acute problems are put forward for further discussion, decision and planning.

The Assumptions

PRA planning as applied to rural development at PRIs level is based on certain assumptions which may be outlined as follows:

- villagers are rational;
- they are capable of generating and cross checking information;
- presence of local NGDO facilitators, elected panchayat and village leaders, key informants and people of the locality is necessary;
- presence of village level workers, elected political leaders (MLA, MP), block and district level officers (specialists) and planning experts;
- people's contribution in the form of cash, kind and labour etc.;
- while planning, involvement of retired personnel in the locality, students from universities and colleges with their professors and local school teachers is required, provided they can contribute both time and talent freely;
- a volunteer group involving people from all sections, government, non-government, retired persons and students could be trained by the state and district planning machinery to check the financial viability, technical feasibility and social acceptability aspect's of the projects.

Planning at Different Levels of PRIs

People's participation in planning process is the main objective of people's plan and development. So at all the four levels, i.e., at village, Village Panchayat, Panchayat Samiti (block) and Zilla Panchayat (district) plans could be prepared subsequently. In PRA people's plan, i.e., planning from below or a 'bottom up' approach is followed for giving people ample opportunity to plan at village level. Village is considered as the most important and basic unit for all the remaining phases, i.e., Gram Panchayat, Panchayat Samiti and Zilla Parishad.

Planning at Village (Gram Sabha)

Village is the most important unit to formulate plan by adopting PRA data base. people's plan at village level ends after passing through three stages subsequently. These are:

Stage I	:	Awareness building for people's plan, collection of information, mapping and diagramming.
Stage II	:	Triangulation and preparation of data base for planning.
Stage III	:	Planning at Gram Sabha/Village Meeting

Stage I

Rural people, NGDO facilitators, PRI elected leaders and village level worker (VLW), key informants and trained volunteers have equal role in the collection of information and presentation on drawing sheets of different problems and resources. In each map and diagram, index and indicators are given for easy understanding. Information relating to the main and related problems can be discussed for better planning and understanding. The following PRA exercises could be taken which would help planning at the village level:

- resource mapping (natural and human);
- social mapping;
- livelihood sources;
- transacts;
- health and sanitation;
- status ranking, scoring and grouping;
- literacy map;
- crop matrix and seasonal diagramming;
- migration and mobility map;
- alcoholism;
- chapati or Venn diagrams;
- historical profile;
- futures possible;
- time trends;
- daily routine diagram; and
- case study.

Stage II

At this stage the maps are prepared carefully on drawing sheets which could be taken up for further discussion and planning. Here each and every map is cross-checked in public view in the Gram Sabha for necessary correction. By doing this a participatory method of learning goes simultaneously with discussion, decision and planning.

Stage III

At this stage in another meeting people try to prioritise their problems and discuss the top priority problems in detail. Both cost-benefit analysis, budgeting and people's contribution are taken care of at this stage, so that people would feel that it is their problem, they have identified it and planned in their own way to solve it. The most important factor is that they contribute in the form of cash, kind or labour to solve the problem.

Planning at Gram Panchayat

After formulation of projects at village level using the PRA data base (primary data), the projects can be sent to Gram Panchayat office. If the secondary data are available at the Gram Panchayat level, they could be used for further discussion. PRA data from all the villages under its purview would make the panchayat office the most reliable resource centre for learning, discussion and planning. On the basis of PRA data and projects prepared at village level on human and natural resource maps, land available and its utilization pattern, health and sanitation, cropping pattern, pros and cons of alcoholism, literacy, well being maps etc., Gram Panchayat level projects can be prepared and sent to block office for approval.

Planning at Block (Panchayat Samiti)

After receiving the formulated projects from different Gram Panchayat offices they are assembled at block office for further cross checking and discussion on their technical feasibility. At this level elected leaders, block level administrative and executive staff, specialists finalise them. Here financial viability factor is also taken into consideration. On the basis of selected projects the budget is prepared and sent to the district or Zilla Parisad for financial assistance.

Planning at District (Zilla Parisad)

The ease at which the district planning can be made by the district planning body and the responsibility it will entail will largely depend upon how they have co-operated and co-ordinated planning at the previous stages i.e., village, panchayat and block. A reliable and authentic data base and facilitation only help to bring out suitable programmes having social acceptability, financial viability and technical feasibility. A trained "volunteer team" including people, specialists, teachers, professors, students, retired government and non-government personnel, development administration, PRIs and NGDO's personnel can be organized with minimum expenditure and basis of planning. The whole process starting from collection of PRA data to formulation of projects and programmes at the district level need long time but facilitation at the district and state planning bodies may minimise it. Above all, awareness building among the people may play a very important role in minimizing the time required for project and programme formulation. At district level there is a chance of bridging the gap between projects and programmes for sustainable rural development. Involvement of rural people in the planning process from below is difficult but not impossible. Kerala has shown the way by keeping forty two per cent of the plan fund for development of the Panchayat bodies. If people will be given power to plan according to their needs then the earlier balance of power with the bureaucracy and legislature, particularly their entrenched interest, may be hampered. So planning from below through people's plan for sustainable rural development is a challenging task in 21st century for all of us.

NOTES AND REFERENCES

1. B. Jain, (1996), *NGOs in Development Perspective* (ed.), New Delhi, Vivek Prakasan, p. 6.
2. Jain L.C., B.V. Krishnamurthy and P.M. Tripathy (1985): *Grass Without Roots: Rural Development Under Government Auspices*, New Delhi, Sage Publications, pp. 50-105.
3. Wolfgang Sachs (1993) *Development Dictionary* (ed.), Johanessburg: Witwateresrand University Press, pp. 126-72
4. S.R. Maheswari (1995) *Rural Development in India—A Public Policy Approach* (Second ed.), New Delhi, Sage Publication, pp. 40-45.

5. P. Jain and T.W. Hochgesang (ed.), (1995): *Emerging Trends in Panchayati Raj in India* (ed.), Hyderabad, NIBR Publication, p. 70.
6. Vimala Ramachandran (1996): NGOs in the Time of Globalization, Seminar 447 – November, p. 57.
7. Neela Mukherjee (1993): *Participatory Rural Appraisal Methodology and Applications,* New Delhi, Concept Publishing Company, p. 30.

5

Strengthening Panchayati Raj Institutions

The Case of West Bengal

DILIP KUMAR GHOSH

The Seventy Third Amendment to the Constitution of India has set in motion a new phase of democratic decentralization through the length and breadth of the country. It not only conferred constitutional recognition to the panchayats but also empowered them for their vibrant growth. Among many changes brought by 73rd Amendment, the enactment entrusted the panchayats with powers, authority and responsibilities (Article 243G). This Article provides that the Legislature of a State may by law endow the panchayats with such powers and authority as may be necessary to enable them to function as institutions of self-government. Such law may contain provisions for the devolution of powers and responsibilities upon the panchayats at different tier for the (a) preparation of plans for economic development and social justice, and (b) the implementation of schemes for economic development and social justice that may be entrusted to them including the matters listed in the Eleventh Schedule of the Constitution. This Eleventh Schedule indicates the arena of activities that the panchayat bodies have to undertake. Along with this provision the 73rd Amendment by Article 243H authorises the panchayats to levy, collect and appropriate such taxes, duties, tolls and fees in accordance with such procedure and subject to such limits as may be decided by the Legislature of a State by law.

These two provisions amply demonstrate that the implications of this Amendment Act 'are far reaching perhaps not fully grasped even by those who legislated the measure' (Mukharji, 1993). But the problem is that the Seventy Third Amendment Act authorizes the State

Legislatures to endow the panchayats with such powers and authority that may be necessary to render function as institutions of self government.

This is the minimum that a state should ensure how far they will go (that is the extent of endowing power and authority to the panchayat bodies) that depends on the state's policy and political will of the ruling party in rein. In fact, the Seventy Third Amendment provided the floor but not the ceiling. "How far above the floor a particular state may go is a question of policy for that state, bearing in mind that the panchayats from now on are to be (a) constitutional bodies, and (b) institutions of self government" (Mukarji & Bandyopadhyay; 1993). The word self-government generally means autonomy, that is, governance without outside interference. But in real situation the existence of perfect autonomy is difficult to be ensured. Because the states and their subordinate units have to function with a framework of federal structure where Indian union has the authority to control and guide the down level units like states, districts and so on. The Seventy Third Amendment Act only constitutionalized the panchayat institutions. To quote:

> "......What the 73rd Amendment has done is to constitutionalize not only the three strata of government but also within the third stratum, the three levels of panchayats. Therefore there will now be a five-storied pyramid of the government, rising from the village panchayats at the base to the union at the apex. At each level below the union there can be no more than an appropriate measure of autonomy. Self-government—at a particular level thus means such partial autonomy as is appropriate for that level. The principle holds for the states as well as for each level of the panchayat" (Mukharji & Bandyopadhyay: 1993, p. 79).

Since the days of Balwant Rai Mehta Committee when it recommended for three tier Panchayati Raj institutions it was there that the local self-government institutions divided authority from the State Governments. The state governments ultimately were the main controller of the fate of these local level institutions. The question is what will be the range of governmental functions that can be assigned to the panchayat bodies and how much autonomy will be allowed to perform these functions. In making solutions to this question the major criterion is to see that the panchayat bodies should get the scope to win the acceptability and confidence of the people through efficient implementation of development programmes, rather than be perceived as organs of state power. For these they should be able to provide responsive administration, which implies a degree of freedom to adjust programmes according to local needs without being hamstrung by detailed government directives. This can be possible when and only when they can mobilize local resources and augment the scale of development effort. The tasks and responsibilities assigned to the panchayat bodies can be ensured if and only if they are strengthened from different angles. Both for the decentralized and participatory development the panchayats in the meantime are accepted as an important vehicle. With this background the present study has been undertaken with reference to West Bengal's experience.

The present study starts with the resource mobilization aspect of the panchayat institution. In West Bengal the mobilization of own resources by the panchayats are very much emphasized towards enhancing the capacities of the panchayat bodies. Secondly, in the decentralized framework for strengthening the panchayats district planning mechanism has been initiated in this state even before the constitutional amendment. This is a process in enhancing the role of the panchayat bodies in the decision making process. Thirdly, the infrastructure building of the panchayats is urgently required for enhancing their strength—this aspect is studied with respect to the arrangements designed by the State Government. Along with the devolution of power to the panchayats the support of the functional departments are very much urgent—the measures taken by the Government of West Bengal is the fourth section of the study. Finally, the strength of the panchayats lies on the people—the provisions made by west Bengal government and the reality are also presented for getting a reflection on the people's involvement in panchayat functions.

Resource Mobilization

In 1962 the Government of India in Ministry of Community Development Panchayati Raj and Cooperation constituted a study team on resources and finances of the Panchayati Raj institutions. The resolution underlying the constitution of the study team (under the chairmanship of K. Santhanam) mentioned that "one of the fundamental requirements of Panchayati Raj is that adequate resources should be available to the Panchayati Raj institutions through transfer or otherwise, to enable them to discharge their responsibilities." According to this resolution "the resources can be broadly classified as follows: resources transferred in the form of grants and loans by Central and State Governments, by higher tiers of Panchayati Raj to the lower tiers, the proceeds of taxes, duties, ceases etc. which the institutions can levy under their own powers, income from remunerative assets developed by the institutions and gifts and donations from public spirited citizens and voluntary contributions by the people".

In setting its approach to the problem of resources and finances of the Panchayati Raj institutions, Santhanam committee elaborated that for performing the functions allowed to PRIs, on one hand, they required 'substantial assistance from Governments' and, on the other hand, for being a self governing institutions' they should have substantial and growing resources'. In the words of the Committee.

> "Financial resources have to be adequate for the functions allotted to each Panchayati Raj institution. Nowhere in the world can local bodies function without substantial assistance from Governments. At the same time it will detract from the dignity and autonomy of a self governing institutions if it is to be entirely dependent on assistance from above. It is essential for stability and growth of these institutions that they should have substantial and growing resources which are entirely within their power to exploit and to develop."

After the Santhanam committee, the major discussion on financial resources of the panchayats was taken up in Asoka Mehta committee report. The chapter on 'Financial Resources' in the report starts like this: "It is generally the view that part of the inability of the Panchayati Raj institutions to come up to the expectation lay in their weak financial resources." In the opinion of the committee there was general reluctance by the Panchayat bodies to impose taxes. This reluctance was not only visible in the grass root tier but also persisted in the upper tiers. To pursue the matter of building up own resources, Asoka Mehta committee recommended that some taxes should be made binding on the panchayats. The committee was fully supportive to the idea of compulsory taxation by the panchayats though in the then situation this was not fully operative. For example, in the states of Assam, Jammu and Kashmir, Manipur, Meghalaya, Orissa, Rajasthan, Sikkim, Tripura and Uttar Pradesh, all the taxes to be imposed were optional for the gram panchayats. However, for enhancing the resource base of the panchayat bodies the committee made a recommendation.

"It is not possible to have a standard list for all states. Certain taxes like house tax, profession tax, entertainment tax, special taxes on land and buildings and certain fees like various market fees should be levied compulsorily by the PRIs at the appropriate level. It is desirable that the State Government itself should specify preferably in the statute itself or in the alternative by rules the minimum and maximum rates of taxes and fees leviable by the Panchayati Raj institutions." (p. 109).

Thereafter, in the 73rd Amendment Act, Article 243H was inserted in the Constitution with the caption "Power to impose taxes by and funds of the panchayats" with the objective to augment their own resources. The provisions enshrined in the Article are (to quote):

"The Legislature of a State may, by law–

(*a*) Authorize a panchayat to levy, collect and appropriate such taxes, duties, tolls and fees in accordance with such procedure and subject to such limits.

(*b*) Assign to a panchayat such taxes, duties, tolls and fees levied and collected by the State Government for such purposes and subject to such conditions and limits.

(*c*) Provide for making such grants-in-aid to the panchayats from the consolidated fund of the State, and

(*d*) Provide for constitution of such funds for crediting all moneys received respectively, by or on behalf of the panchayats and also for the withdrawal of such money therefrom, as may be specified in the law."

In West Bengal Panchayat Act 1973 (as modified subsequently) there are many provisions for augmenting the resource base of the Panchayat bodies at the village, block and district level. At the village level the Gram Panchayat can impose tax and fees on the following items:

(*i*) Tax on land and building at the rate of one percent of the annual value of such lands and buildings when the annual value does not exceed one thousand rupees and at the rate of two percentum of the annual value when it exceeds one thousand rupees.

(*ii*) Additional stamp duty at the rate of two percentum on all transfers of immovable property situated within the jurisdiction of the gram panchayat.

The Gram Panchayats are authorized to levy the following fees, rates and tolls when the maximum rate is prescribed by the State Government:

(*a*) Fees on the registration of vehicles;
(*b*) Fees for providing sanitary arrangements at the places of worship of pilgrimage, fairs and melas in the jurisdiction of the panchayats;
(*c*) Fees on registration for running trade, wholesale or retail unless such registration or trade is prohibited by any law for the time being in force;
(*d*) Fees on licence on dogs and birds and other domestic pet animals;
(*e*) Fees for grazing cattle or grazing land vested in or under the management and control of the gram panchayat;
(*f*) Fees for the use of burning ghat vested in or under the management and control of the gram panchayat;
(*g*) Fees on registration for shallow tube wells fitted with motor-driven pump sets and installed for irrigation for commercial purposes;
(*h*) Water rate may be charged by the gram panchayat where the arrangement for the supply of water for drinking, irrigation or any other purpose is made by the gram panchayat;
(*i*) A gram panchayat may impose a lighting rate where the arrangement for lighting of public streets and places is made by the gram panchayat within its jurisdiction;
(*j*) A conservancy rate may be imposed by a gram panchayat where the arrangements are made for clearing private latrines, urinals and cesspools by the gram panchayats;
(*k*) A drainage rate may be imposed where the arrangement for regular clearance of common drains is made by the gram panchayat;
(*l*) A general sanitary rate may be imposed where arrangement for the construction and maintenance of public latrines is made by the gram panchayat;

(*m*) A gram panchayat may impose toll in respect of any ferry ghat either established by it or under its management.

Tolls at any toll bar on any road or bridge vested in the gram panchayat or under the management of the Gram Panchayat—these tolls can be imposed on persons, vehicles or animals or any class of them.

The panchayat bodies at the intermediate and district level are also authorized to levy tolls, fees and rates. The proceeds from such sources are income of the panchayat bodies. The items on which the Panchayat Samity (intermediate tier) may impose tolls, fees, rates etc. are mentioned in the section 133 of the West Bengal Panchayat Act. In short, these are:

(*i*) Tolls on persons, vehicles or animals at any toll bar or any road other than kutcha road or any bridge vested in it or under its management;
(*ii*) Tolls or ferry established by it or under its management;
(*iii*) Fee for providing sanitary arrangements at fairs, melas, or place of worship or pilgrimage;
(*iv*) Water rate where arrangement for the supply of water for drinking, irrigation or any other purpose is made by the panchayat samity;
(*v*) Lighting rate where arrangement for lighting of public streets and places is made by the panchayat samity;
(*vi*) Licence fee for hut or market;
(*vii*) Licence fee for carrying on any trade or business declared by the State Government to be offensive or dangerous.

Like the Panchayat Samitis, the Zilla Parishads can also levy tolls, fees and rates in the same manner. But Zilla Parishad cannot impose licence fee for hut or market or for carrying a dangerous and offensive trade—this power is given exclusively to the Panchayat Samitis. In addition to the sources directly imposed by the panchayat bodies for their own resource generation, the grants from the State and Union Government are also the important rather, the major source of income. In the WBP Act provisions are made in sections 47A (for gram Panchayat), 135A (for Panchayat Samity) and 182A (for Zilla Parishad) so that the panchayat bodies at the village, intermediate and district level may borrow money from the State Government, or from banks or other financial institutions for furtherance of their objectives on the basis of specific schemes drawn up by them.

The experience of West Bengal on this resource mobilization aspect is more or less dismal. To quote from a publication of the Govt. of West Bengal: "the fact is that the Panchayats are not at all concerned about acquiring even a modicum of financial self reliance... there is near zero self reliance, which means near zero autonomy and correspondingly near zero self government" (Mukharji & Bandyopadhyay 1993). For corroborating this observation in Table 5.1, the collection of taxes by gram panchayats vis-à-vis the assessment of tax collection prepared by them are presented for each district of the West Bengal over the periods 1997-98 and 1998-99.

Table 5.1: Trends of Assessment and Collection of taxes by Gram Panchayats

District	*Number of Gram Panc-hayat*	*1997-98*			*1998-99*		
		A	*B*	*Percen-tage of Collection (%)*	*A*	*B*	*Percen-tage of Collection (%)*
Bankura	190	46.43	15.85	34.14	49.75	19.02	38.23
Birbhum	167	109.94	37.04	33.70	136.11	44.27	32.52
Burdwan	277	259.10	82.55	31.86	262.44	85.02	32.39
Coochbehar	128	72.27	11.11	15.37	73.41	11.43	15.57
Dakshin Dinajpur	65	23.42	11.98	51.15	36.86	11.51	31.23
Darjeeling	104	39.08	16.69	42.71	45.98	18.53	40.30
Hooghly	210	254.57	64.29	25.25	280.97	81.50	29.00
Howrah	157	85.58	31.52	36.83	125.68	39.89	31.74
Jalpaiguri	148	65.69	29.07	44.25	84.58	38.91	46.00
Malda	147	59.22	19.23	32.47	61.23	19.54	31.91
Murshidabad	255	141.77	30.12	21.24	151.04	29.55	19.56
Midnapore	514	1040.61	297.70	28.61	366.34	105.97	28.93
Nadia	187	135.24	39.36	29.10	149.32	54.95	36.80
North 24-Parganas	200	168.64	62.38	37.00	197.15	73.44	37.25
Purulia	170	15.72	1.70	10.81	23.07	1.65	7.15
South 24-Parganas	312	176.32	51.34	29.12	292.13	55.04	18.84
Uttar Dinajpur	99	68.64	11.31	16.48	84.66	11.31	13.36
West Bengal	3340	2762.24	813.24	29.44	2420.72	701.53	28.98

Source: Panchayat and Rural development Department, Govt. of W.B.
Col. A – Total assessment (Rs. in lakhs).
Col. B – Total collection (Rs. in lakhs).

The assessment and collection figures presented in Table 5.1 show that at the state level assessment of taxes was reduced by 12.36 per cent in 1998-99, which is obviously unusual. The collection of taxes was also reduced in 1998-99 by 13.74 per cent. Less collection of taxes simply signifies the unwillingness of the panchayat bodies to collect taxes. At the aggregate level the percentage rate of collection vis-à-vis assessment was also reduced in 1998-99 by 1.56 in comparison to the year 1997-98. In Table 5.2 per capita collection of tax by the gram panchayat in the year 1997-98 and 1998-99 are presented. The estimate of the rural population (district wise) is taken from the Statistical Abstract (1997-98) prepared by the Bureau of Applied Economics and Statistics, Government of West Bengal.

Table 5.2: Rural Population and per capita collection of taxes by Gram Panchayats

	1997-98			*1998-99*		
District	*Rural Popu-lation*	*Total Collection (Rs. in lakhs)*	*Per Capita Collection (in Rs.)*	*Rural Population*	*Total Collection (Rs. in Lakhs)*	*Per Capita Collection (in Rs.)*
Bankura	2842807	15.85	0.56	2890538	19.02	0.66
Birbhum	2620168	37.04	1.41	2672662	44.27	1.66
Burdwan	4492942	82.55	1.84	4594810	85.02	1.85
Coochbehar	2261310	11.11	0.49	2307764	11.43	0.49
Dakshin Dinajpur	1189632	11.98	1.00	1215954	11.51	0.95
Darjeeling	1042772	16.69	1.60	1067923	18.53	1.73
Hooghly	3383737	64.29	1.90	3452913	81.50	2.36
Howrah	2157209	31.52	1.46	2207137	39.89	1.81
Jalpaiguri	2696451	29.07	1.08	2760462	38.91	1.41
Malda	2865517	19.23	0.67	2941201	19.54	0.66
Murshidabad	4928116	30.12	0.61	5052062	29.55	0.58
Midnapore	8528114	297.70	3.49	8710507	105.97	1.22
Nadia	3487699	39.36	1.13	3580282	54.95	1.53
North 24 Pgs	4189200	62.38	1.49	4306128	73.44	1.70
Purulia	2247472	1.70	0.07	2288827	1.65	0.07
South 24 Pgs	5237800	51.34	0.98	5286490	55.04	1.04
Utter Dinajpur	1980814	11.31	0.57	2039328	11.31	0.55
West Bengal	56151760	813.24	1.45	57374988	701.53	1.22

From Table 5.2 it can be seen that in 1997-98 the state average of per capita tax collection by the Gram Panchayats is only Rs. 1.45

as against Rs. 1.22 in 1998-99. In many districts it is even less than a rupee. As a consequence, the gram panchayats have to depend more on the grants-in-aid from the State Government through Zilla Parishad and Panchayat Samiti. This situation can only be altered when "...every Panchayat should aim to raise at least a rupee for every rupee it gets by way of devolution from above. This would double the resources of the Panchayats. What is more, such self raised resources would be wholly united and thus available to be spent in accordance with each panchayat's own priorities." (Mukharji & Bandyopadhyay: 1993).

A Case Study

For ascertaining the situation at panchayat samiti level a quick study was undertaken in the books of account of Karimpur-II panchayat samiti in the district of Nadia. Data in respect of the fourth term of Panchayati Raj in West Bengal (from 1993-94 to 1997-98) was available. In addition, data of the first year of the fifth term of panchayats (1998-99) was also accessed. A Panchayat Samiti in West Bengal has two main sources of getting fund – (*a*) the grant-in-aid from the State Government including schematic fund for implementing different schemes assigned to the Panchayat Samiti, funds covering allowances and honorarium of elected office bearers of the Panchayat Samiti and (*b*) own resources covering tolls, taxes, fees and income through own assets. From Table 5.3(*a*) it can be seen that the share of own resources in the total receipt of fund for the respective year is very nominal. In spite of the provisions incorporated in WBP Act, increased interest is not revealed within the panchayat samiti to build up a sound own fund. For example, in an estimate it is found that there are around 10,000 business units covering all types of business activities – charging Rs. 10 per annum can yield one lakh rupee for the panchayat samiti. But this is not being exploited to the fullest extent. The main collection is coming from own property like ponds, forest, etc. developed through schematic funds.

The year-wise expenditure of the panchayat samiti for the period under the study was taken into consideration to reflect the fact that if the performance be satisfactory, collection of fund from the people will be easier.

Table 5.3(*a*): Total fund received vis-à-vis own resources of Karimpur-II Panchayat Samiti

Year	*Grant-in-Aid from Govt. (Rs. in lakhs)*	*Own Resources (Rs. in lakhs)*	*Total Fund receipt (Rs. in lakhs)*	*Percentage of own Resource in Total Receipt*
1993-94	38.88	0.35	39.23	0.89
1994-95	77.76	0.91	78.67	1.16
1995-96	77.89	0.38	78.27	0.48
1996-97	34.20	0.82	35.02	2.34
1997-98	97.04	2.13	99.17	2.15
1998-99	140.21	1.68	141.89	1.18

Table 5.3(*b*): Extent of expenditure in relation to available fund of Karimpur-II Panchayat Samiti

Year	*Total available Fund (Rs. in lakhs)*	*Total expenditure (Rs. in lakhs)*	*Percentage of expenditure*
1993-94	70.50	55.73	79.05
1994-95	93.44	65.75	70.37
1995-96	105.96	49.28	46.51
1996-97	91.70	72.44	78.99
1997-98	118.43	87.78	74.12
1998-99	172.54	90.42	52.40

In the year 1995-96 the performance of the panchayat samiti in terms of expenditure vis-à-vis total available funds and the own resource position vis-à-vis receipt of fund are the lowest.

District Planning

For the involvement of the panchayat bodies in the decision making process, the Govt. of West Bengal divides the plan provision for each department into two broad sectors – they are state sector schemes and district and below level schemes. The desegregation of outlay is made with the purpose to devolve fund to the district level for necessary decision making there. For extending the process of decentralization this sort of step was initiated in the year 1987-88 when around 39.75 per cent of the total plan outlay was desegregated to the districts. Though there are variations in the share of this outlay over the years (Table 5.4), yet it is the fact that the share is always more than forty per cent. From the very moment of revitalizing the panchayats a system of democratic decentralization was also attempted by the State Government. Well before the 74th Amendment Act in West

Bengal a decentralized planning set up was launched in the district. At the apex there was a District Planning Committee and at the block level Block Planning Committee. As the majority of the members of these two planning committees are the members of the relevant panchayat bodies, the involvement of the panchayat members is very much in decision making process. In the formulation of a block plan, exercises were initiated from the gram panchayat level and block plans thus prepared were integrated with the district plan for formulating District Plan. Till 1994-95 there was a gradual increase in the share of desegregated outlay vis-à-vis total outlay for the respective years (Table 5.4).

Table 5.4.: Disaggregated outlay in Annual Plans: West Bengal

Year	*Plan Outlay (Rs. in lakhs)*	*Amount of Disaggregated outlay (Rs. in lakhs)*	*Percentage*
1987-88	86200.00	34262.39	39.75
1988-89	95088.90	41495.31	43.64
1989-90	111500.00	52984.66	47.52
1990-91	132800.00	65538.91	49.35
1991-92	148600.00	76087.07	51.20
1992-93	150100.00	82234.44	54.78
1993-94	155000.00	91809.00	59.23
1994-95	170600.00	101251.16	61.00
1995-96	220040.00	117583.96	53.44
1996-97	297260.00	153140.61	51.52
1997-98	398134.00	199465.13	50.10
1998-99	459485.00	200899.94	43.72
1999-2000	578725.00	264500.55	45.70

Source: Annual Plans for the different years published by Planning and Development Dept., Govt. of West Bengal

In addition to the desegregation of outlay since 1985-86 a separate budget head was being operated under the nomenclature of 'District Plan Schemes Fund' where the fund was placed solely at the disposal of the District Planning Committees. These funds as envisaged by the Dept. of Planning and Development were mainly meant for meeting the critical gaps between the fund requirements according to the district plan and the available funds from different sources. Presently this system is discontinued. The unique feature of DPS fund was that it was fully untied where decision making rested solely with the panchayat bodies. The non-availability of untied fund and insignificant own resources make the panchayats dependent on the state government. The increasing dependencies lead to erosion in the autonomy of these local bodies. With the passage of the Seventy Fourth-

Amendment Act of the Constitution of India for supplementing the process of devolution of powers to the local bodies, a provision has been made to constitute a District Planning Committee (DPC) in each district of the country. Article 243 ZD of the Constitution of India states that (to quote):

> "There shall be constituted in every state at the district level a district planning committee to consolidate the plans prepared by the panchayats and the municipalities in the district and to prepare a draft development plan for the district as a whole."

The State Legislature has the authority to make laws in respect of the composition, the manner of filling up the seats, the manner of selection the chairperson of DPC etc. According to the Constitutional provisions incorporated in the Article 243 ZD, every district planning committee has to prepare the draft development plan for the district and in the course of preparation of the draft plan the DPCs have to pay regard to the following aspects:

Matters of common interest between the panchayats and the municipalities including spatial planning, sharing of water and other physical and natural resources, the integrated development of infrastructure and environmental conservation. The extent and type of available resources, whether financial or otherwise, and consultation with such institutions and organizations as may be specified by order.

In consonance with the 74th Amendment Act the West Bengal District Planning Committee Act was enacted in the year 1994. This Act was made for the constitution of District Planning Committee at the district level for consolidation of the plans prepared by the Panchayats and the Municipalities in the district and the preparation of a draft development plan for the district as a whole. In WB District Planning Committee Act, three type of members are specified:

- the members to be elected by and from amongst the elected members of the Zilla Parishad, but no Savadhipati (that is the chairperson of the Zilla Parishad) shall be elected as the members according to this provision;
- the members to be elected by and from amongst the elected members of all the municipalities within the district, but no chairman of the municipalities can be the member under this provision; and
- the members to be appointed by the State Government, but

the number of such members shall not exceed one-fifth of the total number of members of the DPC.

According to the provisions made in the WBDPC Act, the State Government will select the members from the following categories of people:

- the Sabhadhipati of the Zilla Parishad;
- the District Magistrate of the district;
- such other members as the State Government may think fit to choose from amongst–
- the *Sabhapatis* (Chairperson) of the Panchayat Samitis (at block level) of the district;
- the members of the House of People and the Legislative Assembly of the State elected from a constituency comprising any part of the district, but they must not be the Ministers;
- the officers of the State Government or of any statutory body or corporation deemed by the State Government to have specialized knowledge;
- the Chairpersons of the municipalities within the district;
- the economist, social and political workers of eminence.

In West Bengal the Sabhadhipati of the Zilla Parishad has been made the chairperson of the DPC and the District Magistrate of the district, the secretary of the DPC. According to the WBDPC Act, the District Planning Committee has been entrusted with the following duties:

- Consolidation of the plans prepared by the panchayat bodies and the municipalities in the district;
- Preparation of a draft development plan for the district as a whole;
- In course of preparing the draft development plan paying attention to the subjects like matters of common interest between the panchayats and the municipalities including spatial planning, sharing of water and other physical and natural resources, the integrated development of infrastructure and conservation of environment; and
- Taking account of the extent and type of available resources, whether financial or otherwise.

On the basis of the provision of the WBDPC act, the State Government made the West Bengal District Planning Committee Rules, 1994 for carrying out the purposes of the DPC act. In the Rule the provision has been made to constitute (a) Urban Development Sub-committee, (b) District Planning and Co-ordination Sub-Committee. As suggested by the Rule the sub-committee meetings are to be held as frequently as necessary but at least once in every three months. For such meetings, no quorum is necessary and the sub-committee shall devise its own procedure for functioning. The District Planning and Co-ordination sub-committee according to the rule shall advise and assist the District Planning Committee in the preparation of the district plan, co-ordinating and consolidating the annual plans and programmes of the municipalities, panchayats, other development authorities and government departments functioning within the district planning area. In addition to these two committees, there is a provision that the State Government may constitute many other sub-committees by notifications.

The West Bengal District Planning Rule specifies that within one month from the date of adoption of the draft development plan by the DPC the Chairperson of the Committee or the secretary of the Committee will forward the draft plan to the State Government (in Planning and Development Department). The State Government then in consultation with the State Planning Board and after giving DPC an opportunity to present the plan shall extend its approval to the plan with any modification if the necessity is felt by the State Govt.

The formation of DPCs in all the districts of the state widens the scope in favour of the panchayat bodies to concretize the decision-making process. "The entire process begins with open mass meeting at the village level for prioritization of plan schemes and then these plans are appropriately integrated at the Gram Panchayat level, the Panchayat Samity level and finally at the District Planning Committee level with involvement of the Zilla Parishad". (from Forward by Finance Minister of West Bengal in Annual Plan 1998-99).

At the national level the Ninth Five Year Plan document emphasizes that the establishment of DPC should receive the top priority and should be in a position to formulate local level plans. The Ninth Plan in an unflinching voice noted that 'what can be done at the lower level should be done at that level and not at the higher level'.

The DPCs according to the Ninth plan Document "should not only consolidate plans from below but should take decisions on the development of the district within the given resource potential and identified needs and constraints". The emphasis on decentralizing the decision making process will lead to increase the status of the panchayats as without their unflinching involvement plans cannot be designed at all.

Infrastructure Building of Panchayats

With the funds provided by the Tenth Finance Commission, the Govt. of West Bengal initiated the infrastructure building of the panchayat institutions throughout the state for enhancing the efficiency of the panchayat offices in discharging their responsibilities. Initially priority was given to the Gram Panchayats. The State Government directives in this regard are very specific. To quote:

> "Funds sanctioned should be utilized primarily for the implementation of schemes for development of infrastructure of Gram Panchayats only. Schemes for infrastructural development of Panchayat Samitis as also schemes for creation of remunerative assets for any panchayat shall not be taken up for implementation at the initial stage without specific permission of this Department [Panchayat and Rural Development Dept.]" (the order was issued in May, 1997).

For the Gram Panchayats, the State Government provided an illustrative list of schemes for infrastructure building wherefrom the gram panchayats would identify their need for fulfilling the gap. The list of the schemes suggested is of the following nature:

- the purchase of bicycles (not exceeding three per gram panchayats) which will help in making communication easier at the village level as these are the widely used mode of conveyance in rural areas;
- purchase of typewriter (preferably Bengali script), duplicating machine, steel almirah and any other office equipment that may increase the efficiency;
- construction/repair/extension of Gram Panchayat office Buildings (Panchayat Ghar) whichever is needed;

- schemes for construction of link road from the main road to the gram panchayat office, if any bridges, culverts are necessary on such link road that too can be taken up;
- electrification, installation of telephone etc. at the Gram Panchayat office buildings in case of their absence;
- construction of toilets, particularly for the lady members;
- construction of a permanent source of drinking water;
- construction of boundary wall if it is not there.

The State Government in Panchayat and Rural Development Department has taken the decision to allocate the Tenth Finance Commission funds (Rs. 333.45 crores) within a span of four years commencing from 1996-97. The entitlement of the district (three tier panchayat bodies of the district taken together) out of this grant has been worked out on the basis of the recommendation of the State Finance Commission. The fund for the panchayat bodies are being routed through the district administration and are allotted in eight half-yearly instalments. In Table 5.5 the entitlement of the Tenth Finance Commission Fund for the district of the state is presented. The Government of West Bengal directives mention a condition recommended by the Tenth Finance Commission regarding entitlement to this fund. In its recommendation, the Tenth Finance Commission laid down, interalia, that the local bodies should be required to provide suitable matching contributions by raising resources. Though the Tenth Finance Commission report did not specify any quantum of contribution or proportion *vis-à-vis* total grant of the panchayats, the State Government in its directive mentioned that:

> "... pending final decision (of the Government of India), it is made compulsory that each Gram Panchayat for the present should contribute a share of a least ten per cent of the amount placed with it, as its matching contribution. This contribution should not be raised from any schematic or establishment grant received from the Central or State Government. The fund should come out of its own resources in true sense of the term..."

This conditionality was imposed with a view to encouraging collection of own resources by the panchayats.

Table 5.5: Entitlement of Tenth Finance Commission Fund to the District of West Bengal

District	*Entitlement (Rupees in cores)*
Bankura	16.839
Birbhum	15.148
Burdwan	27.796
Cooch Behar	14.792
Dakshin Dinajpur	10.163
Darjeeling	6.973
Hooghly	19.147
Howrah	15.009
Jalpaiguri	17.046
Malda	16.385
Midnapur	40.301
Murshidabad	26.499
Nadia	19.387
North 24 Parganas	30.000
Purulia	14.73
South 24 Parganas	30.424
Uttar Dinajpur	13.468
West Bengal	333.450

Source: Panchayat and Rural Development Department, Government of West Bengal.

For monitoring the utilization of funds placed for the infrastructure building of the panchayats the State Government constituted sub-committees for each district. According to the Government's instruction the sub-committee includes Sabhadhipati of the Zilla Parishad as Chairman, the District Magistrate of the district as member and the Additional Executive Officer of the district as the Secretary of the sub-Committee. This sub-committee has the authority to co-opt any other elected member of the Zilla Parishad subject to the condition that the total number of such co-opted members should not exceed two. This sub-committee has the following functions:

- The committee considers the proposals of infrastructure development submitted by the gram panchayats and panchayat samitis of the district, extends the approval on the basis of the guidelines and government order set for this scheme;
- Recommends the release of fund to the Zilla Parishad in favour of the concerned gram panchayat/panchayat samiti;
- Undertakes monitoring of the schemes taken up for execution.

As regards the utilization of fund under this head, the status of the districts at the end of the financial year 1999-2000 is reflected in the Table 5.6.

Table 5.6: Allocation and expenditure of fund In Infrastructure Building of Panchayats

District	*Allocation of Fund (Rs. in crores)*	*Expenditure (Rs. in crores)*	*Percentage of expenditure*
Bankura	11.881	2.311	19.44
Birbhum	10.688	6.487	60.71
Burdwan	19.612	7.754	39.54
Cooch Behar	10.437	4.623	44.29
Dakshin Dinajpur	7.173	5.411	75.43
Darjeeling	4.920	2.115	42.98
Hooghly	13.510	5.002	37.02
Howrah	10.588	4.294	40.55
Jalpaiguri	12.026	5.321	44.24
Malda	11.562	5.086	43.98
Midnapore	28.435	8.852	31.13
Murshidabad	18.697	9.270	49.58
Nadia	13.680	5.527	40.40
North 24-Parganas	21.170	8.575	40.50
Purulia	9.928	3.971	39.99
South 24-Parganas	21.466	7.836	36.50
Uttar Dinajpur	9.502	4.502	47.38
West Bengal	235.275	96.937	41.20

Source: Panchayat and Rural Development Department, Government of West Bengal

From Tables 5.5 and 5.6 it can be noticed that of the fund entitled for the district, 70.56 per cent were allotted till the end of March 2000. Of the fund thus allotted the expenditure at the state level is to the extent of 41.20 per cent only. There are wide variations within the districts – from 19.44 per cent in Bankura to 75.43 per cent in Dakshin Dinajpur.

In addition to the infrastructure building of the panchayats, the Government of West Bengal is channelising the funds from Tenth Finance Commission to the panchayat bodies for the creation of the remunerative assets. The instruction from the state government specifies the nature of the assets to be created and the expected contribution from such assets. According to the government instructions the assets will be of the following nature:

- the assets must be the source for one time or periodical income for the panchayat bodies;

- any proposal for asset generation should be within the functional area of the panchayat body as laid down in the West Bengal Panchayat Act;
- the assets thus created may open job opportunities for the private entrepreneurs but this must not lead to the creation of any post any where in the panchayat bodies;
- the maintenance of such assets may be done on contract or lease agreement basis;
- the assets shall not have any dependence on any commercial activity involving fluctuations in market prices;
- the assets shall not be based on real estate business;
- the assets will add to the socio-economic development of the area and welfare of the weaker sections, on one hand, and income generating, on the other.

The illustrative list includes construction of small bridges on rivers/ rivulets (for which tolls may be collected through some engaged agency), construction of market complex, bus/truck stand, sulabh complex (urinals and latrines) in market places etc. The main intention of this scheme for the creation of Remunerative Assets and other Development Programmes in Panchayat bodies is to give sufficient annual income to a panchayat body so as to cover interest on investment and return of the capital within reasonable period, for example, say, within 10 years as suggested in the government directives. With the remunerative assets, thus formed, it is expected that the panchayat bodies will solidify their financial aspects.

Transfer of Personnel and Administrative Powers to the Panchayat Bodies

According to the provisions made in the West Bengal Panchayat Act, 1973 the State Government may place at the disposal of the panchayat bodies the services of any officers or employees serving under it on such terms and conditions as the State Government may think fit. At the same time there are provisions in the West Bengal Panchayat Act to recall the services of such officers or employees when the gram panchayat or panchayat samiti or zilla parishad passed a resolution to that effect at a meeting specially convened for the purpose by a majority of the total number of members holding office for the time being.

In the Standing Committees at the Panchayat Samiti and Zilla Parishad level, the State Government, according to the provisions of the WBP Act [section 124 and 171 and their sub sections 2(c)] may appoint such number of persons being officers of the State Government or of any statutory body or corporation or being eminent persons having specialized knowledge as the State Government may think fit. These provisions are made with a view to extending support to the panchayat bodies in their functioning. In the process the panchayat bodies are being strengthened for discharging the duties and responsibilities entrusted to them. In addition to these two provisions, according to section 125, sub-section 4(c) and section 172, sub-section 4(c) the Standing Committees of both the panchayat samiti and Zilla Parishad can call for the attendance of any officer in the meeting of the respective Standing Committee. Therefore, in WBP Act the provisions are very much specific in respect of getting the services of any officials of the state government for furtherance of the panchayat activities. However, progress in this respect is very limited.

By a resolution in the year 1997 the Department of Forest, Govt. of West Bengal, has taken a decision to place the services of the departmental officers to the panchayat bodies at the different tiers. The Resolution (dated May 29, 1997) clearly noted that for the successful implementation of the different schemes under the forestry sector as it largely depends on the technical support and specialized knowledge of the officials of the forest department, on one hand, and on support of the local leadership with intimate knowledge of local conditions and preferences ensuring local participation, on the other hand, it is imperative to maintain a close relationship between this department's officials and the representative of the panchayat bodies at the different tiers. All the forest officials starting from the Divisional Forest Officer (DFO) down to the Beat Officer at the block level have been linked with the corresponding panchayat bodies by giving a panchayat – related designation. For example, the Divisional Forest Officer has been designated as the Divisional Forest Officer and ex-officio Parishad Forest Officer as DFO is corresponding to the district level. Similarly, the Range Officers are designated as the Range Officers and ex-officio Samiti Range Officer and the Beat Officers as Beat Officer and ex-officio Samiti Beat Officer as both the Range Officer and Beat Officer correspond to the Block level (i.e., the intermediate tier of the panchayat set-up). The justification of giving such Ex-officio designation has also been given in the Resolution under reference. The Resolution mentioned that 'by virtue of such additional Ex-officio designation, the officer concerned in addition to his own duties shall be responsible concurrently to the Panchayat for assisting them in execution of schemes under forestry sectors as may be entrusted to them from time to time'.

The Department of Health and Family Welfare, Govt. of West Bengal, in a memorandum framed in the year 1995 developed a manner for integrating the health system in the panchayat set up. The Department felt that (as noted in 1995 Memorandum) "while the technical aspects of the Medical Administration may continue to be looked after by the experts, all other areas requiring popular participation should have the benefit of the leadership of the panchayats". The Deptartment of Health and Family Welfare formed different committees at the sub-Centre level, Block level, sub-Divisional level and the District level for monitoring the programmes of the department. Except the sub-Divisional level, all committees are headed by a panchayat representative. For example, the sub-Centre level committee is headed by the Pradhan of the Gram Panchayat where the sub-Centre is located; the Sabhapati, Panchayat Samiti is the Chairperson of the block level monitoring committee while the Sabhadhipati of the Zilla Parishad is the Chairperson of the district committee. As there is no corresponding panchayat body at the sub-Division level, the sub Divisional Officer has been made the chairperson of the sub-divisional Monitoring Committee. In the same memorandum the panchayat samiti has been authorized to adjust the sector areas (area of operation) of sub-Centre so that no sub-centre area falls within the jurisdiction of more than one gram panchayat.

The Department of Health and Family Welfare delegated the power of the registration of births and deaths under the Registration of Births and Deaths Act, 1969 to the Gram Panchayat. The Panchayats are authorized to select the staff support for the specific job from among the health functionaries working in the area.

In the primary education sector the School Education Department (Primary branch) of the Govt. of West Bengal constitutes a Village Education Committee (VEC) for each village in the districts covered under the District Primary Education Programme (DPEP). The main purpose of the Committee is to assist in implementing DPEP. The Pradhan/Upaprodhan or the senior most Gram Panchayat member who resides in the village concerned is the Chairperson of that VEC. Initially the VECs are formed in the year 1995. The Panchayat Samity (at the block level) is authorized to constitute the VEC at the village level. In the VEC, head teacher of the primary school of the village, member of Panchayat Samiti/ Zilla Parishad if resides in that village, anganwadi workers, librarian of the rural library, representatives from sub-Inspector of schools, guardians, representatives from women, SC/ST and minority are included. The panchayats are authorized to lead the team in implementing the scheme. The Panchayat Samiti is authorized by the State Government to monitor and evaluate the work of VEGs in their jurisdiction. If situation demands the Panchayat Samiti can even reconstitute a VEC, then the tenure of these reconstituted VEC will be for the remainder of the term of the Panchayat Samiti as the term of VEC will be the same as that of the Panchayat Samiti'. The VEC alongwith many other tasks assigned by the state government has to arrange additional resource requirements and to ensure convergence of different rural services provided by the government. The constitution of VEC encourages integration of the people and panchayat leadership in the matter of enhancement of the primary education at the village level. In this way the Government of West Bengal is making efforts to integrate the official system with the panchayat network for reinforcing the panchayat institutions so that people can take the lead in the development process and development will be people-centered.

Making Panchayat Participatory

Though the provisions of the West Bengal Panchayat Act in respect of the participation of people in the activities of the panchayats is very specific and well noted, the reality is not that much encouraging. The system of convening two meetings of Gram Sansads introduced in the year 1994 by the West Bengal Panchayat (Amendment) Act, 1994, got momentum only from the year 1996. Table 5.7 is constructed on the basis of the percentage of Gram Sansad meetings held in a district vis-à-vis the total number of Gram Sansads in that district for the period May 1996 to November 1999.

From Table 5.7 it can be seen that 100 per cent coverage of the Gram Sansad meetings are few and far between. For example, since May 1996 to May 1998 (when the Fifth panchayat general election was held) 100 per cent Gram Sansad meetings occurred only 5 times – once in Bankura (November 1997), twice in Coochbehar (November 1996 and May 1998), once in Jalpaiguri (November 1997) and once in Uttar Dinajpur (November 1997). Since November 1998 to

Table 5.7: Percentage of Gram Sansads where meetings were held

District	*May 1995*	*Nov. 1996*	*May 1997*	*Nov. 1997*	*May 1998*	*Nov. 1998*	*May 1999*	*Nov. 1999*
Bankura	44	61	81	100	48	98	85	87
Birbhum	40	95	96	96	20	100	99	96
Burdwan	69	95	87	86	31	100	100	97
Coochbehar	57	100	99	99	100	100	100	99
Dakshin Dinajpur	58	99	88	97	35	100	100	100
Darjeeling	57	55	40	95	74	100	100	100
Hooghly	34	91	73	95	9	99	95	83
Howrah	40	56	80	80	7	100	100	97
Jalpaiguri	43	93	71	100	52	95	95	99
Malda	47	77	70	91	48	100	98	93
Midnapore	84	98	85	99	38	98	97	98
Murshidabad	44	73	57	76	40	100	95	96
Nadia	77	98	90	99	81	100	97	97
North 24-Parganad	25	87	85	99	45	100	98	99
Purulia	30	85	70	92	43	100	98	95
South 24-Parganas	48	83	60	50	5	99	99	99
Uttar Dinajpur	63	96	79	100	16	99	100	100

November 1999, 100 per cent convening of the Gram Sansad meetings occurred eighteen times – the district-wise number of occurrences are given below:

District	*Number of Times*	*District*	*Number of Times*
Birbhum	1	Malda	1
Burdwan	2	Murshidabad	1
Coochbehar	2	Nadia	1
Dakshin Dinajpur	2	North 24-Parganas	1
Darjeeling	2	Purulia	1
Howrah	2	Uttar Dinajpur	2

Only 18, 100 per cent meetings means the achievement is to the tune of 35.29 per cent as within the span of the period from November 1998 to November 1999 three meetings of Gram Sansad can be held in each district. Till the period May 1998 in the state as a whole there are 36217 Gram Sansads in 3314 Gram Panchayats. In the Fifth panchayat general election the number of Gram Sansads in the state was increased because the representation of people per Gram Panchayat member was extended. The number of Gram Sansads increased from 36217 to 44506, i.e., an enhancement of 22.89 per cent. Through continuous persuasion from the state government it became possible to improve convening of meeting after May 1998 election. The state scenario is summed up below for getting an overview of the status of the Gram Sansad meeting.

Period of Meeting	*Percentage of Gram Sansad meetings held*
May, 1996	42.58
November, 1996	86.86
May, 1997	77.42
November, 1997	88.84
May, 1998	28.81
November, 1998	99.10
May, 1999	97.39
November, 1999	96.27

The overall picture of the state thus demonstrates that though the Gram Sandads are accepted as the basic unit of democracy and a forum for interaction with the common masses to ensure transparency, in sizeable cases the meetings are simply not convened – this aspect gives signal of non-participatory character of the panchayats.

Another platform of interaction with the common people is the meeting of the Gram Sabha where the people have the scope to have the deliberation with the panchayat people in respect of their activities. In spite of the best efforts on the part of the state government, the success on this front so far is also limited. Table 5.8 presents the district-

Table 5.8: Percentage of Gram Sabhas where meetings were held

District	*1996*	*1997*	*1998*	*1999*
Bankura	62	86	24	92
Birbhum	86	92	97	70
Burdwan	67	68	100	87
Coochbehar	100	100	99	85
Dakshin Dinajpur	78	86	100	100
Darjeeling	53	56	94	92
Hooghly	61	80	87	67
Howrah	50	62	94	95
Jalpaiguri	75	82	68	77
Malda	67	71	96	97
Midnapore	61	76	70	54
Murshidabad	73	77	90	85
Nadia	91	95	97	96
North 24-Parganas	41	77	94	92
Purulia	62	68	78	86
South 24-Parganas	63	67	95	72
Uttar Dinajpur	83	87	99	88

wise percentage of Gram Sabha meetings held since the year 1996. One point is that though the Government of West Bengal through the amendment of WBP Act introduced section 16B for obligatory holding of the Gram Sabha meeting, in the year 1995 only 8 districts confirmed that the meetings were held.

Hundred per cent coverage of Gram Sabha meeting was possible only in case of two districts – the districts are Coochbehar (for the year 1996 and 1997) and Dakshin Dinajpur (for the year 1998 and 1999). Thus, in case of 68 occurrences (17 district multiplied by 4 years), hundred per cent coverage was possible in case of 4 only – in percentage, 5.88 per cent. Therefore the State Government venture to make the panchayats fully participatory is yet to gain significant results. It becomes necessary to consider similar provisions for the Panchayat Samitis and Zilla Prishads to convene public meetings (like Gram Sansad and Gram Sabha) at least once in a year. So far in WBT Act there is no such provision. The public meetings and public interaction always strengthen the people's body.

The discussion on the process of strengthening panchayat bodies will not be completed without having a glance at the importance of training input. Because training is a vital input in the process of capacity building of the panchayat members who are coming from diverse socio-economic situation with educational attainment varying from illiteracy to graduation and above. To deal with such a situation the Govt. of West Bengal made an elaborate arrangement for training the panchayat members at all levels. In the year 1999-2000, Rs. 253.10 lakh was allotted to the district authorities for imparting five days' training to the panchayat members (Annexure I). After 1998 panchayat election in West Bengal there are 49225 Gram Panchayat members, 8516 panchayat samiti members and 716 Zilla Parishad members (all these figures excluded Darjeeling district where panchayat election was not held in the year 1998). Thus, arrangements were to be designed to train 58457 members. Keeping in view this large number the State Government made directives to form a district level training team for this purpose. This district level training team is being trained at the State Institute of Panchayats and Rural Development (SIPRD) through Training of Trainers. In this way training becomes decentralized and an intermediary group is formed in between SIPRD and panchayat members. However, for Zilla Prishad and Panchayat Samiti officials direct training from SIPRD are organised. Training modules are designed in such a fashion that the programme can impart both knowledge and skill to the participants.

Conclusion

With the amendment of the Constitution of India for extending the Constitutional status to the panchayat institutions the task of strengthening them from both financial and administrative points of view are very much required. But the progress throughout the country is not uniform. The panchayat bodies, in particular at the grassroots and intermediate level in more or less all the states, are financially weak and starving from funds. A long list of duties and responsibilities are made for them but required quantum of funds for discharging those responsibilities is not pouring into. Even the devolution of powers to the Panchayats is sporadic. Some states like Kerala have greater achievement than the other states. Due to this unequal development in the process of empowering the panchayats, the Government of India itself developed a mechanism through development of a Panchayati Raj Devolution Index. Upto June 1999, only in nine states, District Planning committees were constituted though such formation was mandatory according to the 74th Amendment to the Constitution. The states were Kerala, Madhya Pradesh, Rajasthan, Sikkim, Tamilnadu, Tripura, West Bengal, Haryana (in only three districts), Karnataka (in only 10 out of 20 districts) and two Union Territories, namely Andaman and Nicobar Islands and Daman and Diu (source: Panchayati Raj update, June 1999). In the capacity building process, alongwith devolution of power and financial strengthening training is also important. All the members of the three-tier panchayats need this training whose focus area in fact concerns with both the administrative and programme knowledge, on one side, and skill, development and attitudinal change, on the other side. The task is gigantic considering the number of members to be trained. But in view of the fact that the panchayats now become the institutions of self-government, their strengthening is very much essential for the cause of the democracy.

REFERENCES

Mukharji, Nirmal & Bandyopadhyay, D., "New Horizons for West Bengal's Panchayats" published by Govt. of West Bengal, Dept. of Panchayats, October, 1993.

Government of India:

—The Study Team on Resources and Finances of Panchati Raj Institutions (Santhanam Committee), 1962.

— The Committee on Panchayati Raj Institutions, 1977 (Ashok Mehta).
Planning Commission, Government of India, Ninth Five Year Plan: 1997-2002, Vol. 1, 1999
Government of West Bengal, Planning & Development Department:
Annual Plan, 1998-99
Annual Plan, 1995-96

ANNEXURE 1

District wise allotment of funds for training of panchayat members: 1999-2000

District	*Funds Allotted (Rs. in lakhs)*
Bankura	15.80
Birbhum	18.00
Burdwan	18.50
Coochbehar	3.80
Dakshin Dinajpur	7.90
Hooghly	19.00
Howrah	15.30
Jalpaiguri	14.80
Malda	11.00
Midnapur	43.00
Murshidabad	10.50
Nadia	23.00
North 24-Parganas	4.00
Purulia	11.50
South 24-Parganas	29.00
Uttar Dinajpur	8.00
Total	**253.10**

6

Participatory Micro-Planning for Village Eco-Development

AMIT HAZRA and JAYANTA CHOWDHURY

Degradation of forest resources poses serious problems for human-environmental relations in India. Ten million people live within or adjacent to forest land and depend on forest resources for their survival. The socio-economic and cultural system of these people are inter-woven with the ecology of the forest, and if the forest is degraded, life of forest community would be at stake. Side by side, it could become a pathetic situation for wild animals also. However, while forest communities are part of the deforestation problem, they are also a major asset in finding solutions to the problem.

This paper is the output of Social Assessment study of different village eco-development projects of Protected Areas (PAs) of Buxa Tiger Reserve of West Bengal. It is segmented into three sections. Section I deals with some conceptual ideologies related to this topic. Practical experiences about the participatory micro planning have been discussed under Section II, while Section III concludes with some problems and a solution model for solving these problems.

I

Conceptual Issues

The word 'micro planning' refers to the blueprint of action addressing particular objective in a particular area. It has to be qualified with communities' participation. Together it provides a perspective on which development could be properly understood. Below we discuss some questions regarding communities' participation for a clear understanding of the issues related to participatory micro-planning.

What does Participation mean?

Participation means that people are closely involved in the economic, social, cultural and political processes that affect their lives (UNDP).

According to Paul (1987) "Communities' participation' is an active process by which beneficiary or client groups influence the direction and execution of a development project with a view to enhancing their well-being in terms of income, personal group self-reliance or other values they cherish."

Thus, participation *in development* means how community members can be assured of the opportunity of contributing to the creation of the community's goods and services.

But participation cannot be imposed on the people from above. It should be voluntary and based on *will to participate.* It is a spontaneous process not superimposed by higher authority. Until or unless they are well aware about their interest and importance in a particular action, no 'magical' involvement will be found. Here participation means direct involvement of people and not indirect involvement through their representation.

Why Participation?

Community participation, being one of the strong features of decentralized planning, is important to achieve socio-economic uplift for all sections of people because of the following reasons:

(*i*) Active involvement of the community helps in the preparation and execution of effective development action plans by making assessment of the felt needs and constraints of the people easier.

(*ii*) Participation of community members in the development planning process helps in mobilizing resources for effective plan implementation. Besides, it also minimizes the cost of implementation of the developmental programme by reducing the cost of supervision and by eliminating irrelevant components in the action plan. Apart from these, people's participation steps up the speed of implementation of the programmes by mobilizing popular support and cooperation between groups having diversified objectives and interests.

(*iii*) Community participation in developmental activities makes the tasks of monitoring and supervision more effective. This ultimately reduces the amount of leakages of resources, both material and human, in the delivery system and in this process minimizes the "distortion effect" in the implementation of the plan.

(*iv*) Involvement of people in the process of development planning also acts as a "pressure group" from below, thereby creating conducive environments for formulation of effective plans and its better implementation.

(*v*) Finally, the most important agenda for community participation in the process of development, plan formulation and its implementation is that it is primarily required to reduce unequal distribution of power and means of production between groups so as to "positively restructure" our existing society in favour of the so far deprived sections, since generation of poverty is a function of concentration of power in a few privileged class or groups in our society and poverty acts as a major obstacle to achieve the objective of development effort.

Whose Participation?

It is a crucial question: who will participate or whose participation is essential for regeneration? For this it is necessary to identify the stakeholders of that project. Stakeholder is any person who is directly related to the project or group or institution and who will act as the interest group for the project. Thus, it includes intended beneficiaries, intermediaries, witnesses and losers. According to their significance stakeholders can be categorized into two groups:

1. *Primary Stakeholders*: Who are expecting to get benefit from or to be adversely affected by the project intervention.
2. *Secondary Stakeholders*: Those with some intermediary roles.

The multiple uses of Protected Area (PA) resources and the long history of human habitation in forest areas are reflected in the variety and number of stakeholders found in the project sites at Buxa Tiger Reserve (BTR). In this project site the primary stakeholders are mainly those traditional livelihoods who are highly dependent on access to forest resources and the secondary stakeholders are Government agencies operating in and around the PA in collaboration with NGOs and researchers. Now it is clear why the tribes or non-tribes are assumed to play an active role in the making of micro-plan along with the concerned government authority and reputed NGOs.

Participation to What Extent?

Community people must be involved in the consequent activities of project formulation, which affect every stage of a project. They should participate in the—

(*i*) decision making (problem identification)
(*ii*) planning and implementation
(*iii*) monitoring and evaluation
(*iv*) sharing the benefits of development

So people must be at the centre of development. Development has to rotate around people, not people should dance surrounding the development.

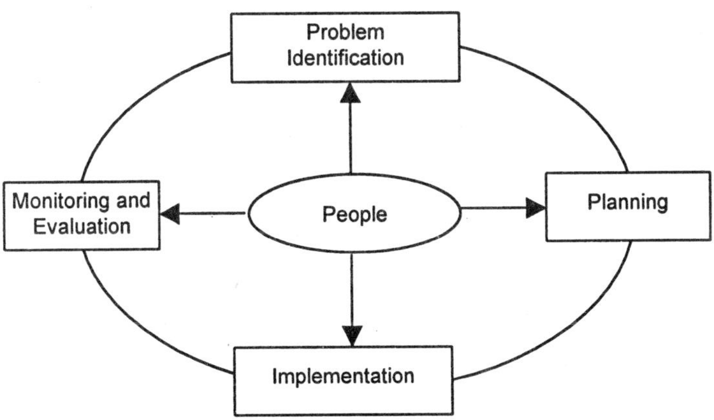

Fig. 6.1: Role of People in Participation

Participation in Which Way?

Participation of people cannot be understood without its proper context. As every task involving people has various dimensions, participation of people has to be visualized in the context of different aspects of developmental tasks. They could be as follows:

Participation as Contribution

The dominant interpretation of participation in development projects in the Third World sees participation as implying voluntary or other forms of contributions by rural people to pre-determined programmes and projects. It is indeed the fundamental tool for success. There are, of course, a whole variety of ways whereby these contributions are forthcoming and managed but, whatever the guise under which they are presented, they form the core of the participatory element in the project.

Participation as Organization

There has been an argument across the range of development literature and practice that organization is a fundamental instrument of participation. The distinction lies between the origins of the organizational form which will serve as the vehicle for participation.

Participation as Empowerment

The notion of participation as an exercise of empowering rural people has gained wide support. Some see empowering as the development of skills and abilities to enable rural people to manage better, to have a say in or negotiate with existing development delivery systems, while others see it as more fundamental and essentially concerned with enabling rural people to decide upon and to take the actions which they believe are essential to their development. Whatever the disagreements in perspective, the relationship between participation and power is now widely recognized.

II

Empirical Eco-Development

During Social Assessment Study of Joint Forest Management (JFM) in the protected area (PA) of Buxa Tiger Reserve (BTR), we have experienced a marvellous example of people's participation to protect the forest as well as its habitats and themselves.

The susceptibility of biodiversity conservation in protected areas (PAs) depends upon the nature of human interactions with PA. The basic design of the eco-development project is based on this understanding. The establishment of P.A. has a significant effect on a variety of human population groups. These groups also have significant impact on the P.A. The eco-development project was designed to minimize the negative impact and strengthen the positive outcomes. The innovative participatory approach for micro planning by the project planners would enable each of these investments to be designed and negotiated by each community backed by the PA authorities as per the felt needs of the specific sites and population.

The following innovative approaches accentuated people's participation for micro planning towards better protection and conservation of biodiversity in the BTR.

Frequent Meeting

From the very beginning the government officials and NGOs interacted with the villagers and discussed with them different topics related to the protection of the forest. In this way a general awareness regarding the forest has been conceived and through different mutual understandings and sharing awareness increased and conflicts reduced. Ultimately Forest Protection Committees (FPCs) have been formed thereafter and were approved and registered by the Forest Department.

Organization Formation

At the village level complete site specific village Eco-Development Committees (EDCs) have been formed. These EDCs are supposed to reduce negative interactions of local people on biodiversity and increase collaboration of local people in conservation of biodiversity. These groups were generally constituted in a small and homogeneous village hamlets and called *user groups,* in which all households are represented in the decision making irrespective of men and women members of the households.

Local Capacity Building

Today it is recognized by all that rural people possess adequate knowledge about the existing local resources and their felt needs, which could be the asset to the developmental agencies. A micro-planning support team composed of PA personnel, collaborating with EDCs/FPCs be formed to develop and implement site specific reciprocal plans on a continuing basis throughout the project period in order to ensure that these plans should meet project objectives and embody the active commitment and participation of the local people.

Accountability and Transparency

These are the most crucial indicators of good governance and decentralized planning. Each FPC in Buxa Tiger Reserve (BTR) maintains a Bank account by the committee. It ensures audits of all fund accounts, including EDC accounts and inspects EDC accounts when required. All the members and concerned authorities are well aware about the balance sheet of the accounts. As the villagers handle all the resources, it ensures the liability and responsibility of the project and also enhances people's interest and involvement.

Compulsory Contribution

One of the most important features of village eco-development project is that local people will contribute at least 25 per cent of the total costs of the project in the form of cash and/or kind (labour) and that payment for the market value of contributions be made available to the local people for additional community investment. The history of rural and urban development investments has shown that where the objective of an investment is to require some form of cost sharing, the projects are highly likely to fail. Thus, if the village eco-development project fails, local people would not continue to commit to the action that improves conservation.

But one should understand why this contribution is essential. These contributions are effective in two ways. Through this type cost sharing people become serious and get more interested to success of the project as they think it as their own project. In another way, this contribution may be reinvested for another activities.

Empowerment to the Vulnerable Stakeholders

Tribals: A variety of tribal groups reside in and around the project sites (at least 44 per cent). Although they pursue a range of livelihood strategies, tribals in general have cultural, social, and economic system closely linked to their environment. So it is not only necessary but also imperative to make sure the participation of these traditional forest dwellers in the project to conserve the biodiversity for their own sake.

Scheduled Castes: Another key group of vulnerable stakeholders in PA sites includes scheduled castes who are essentially poor people with limited ability to generate substantial income. As a group they are socially marginalized, which limits their ability to improve livelihoods without the benefit of the variety of government programme specifically developed to support them. The category of poor people would typically include, in addition to scheduled castes, the landless, sharecroppers and marginal cultivators for whom wage labour is an important source of income. Inside the protected area of BTR, there are 36 per cent of SC people. So, for making a successful micro-plan it requires to include these people for better result.

Women: As a group, women are perhaps the most marginalized and vulnerable among the stakeholders in the project sites, particularly those belonging to the ST and SC communities. Women are for the most part primary collectors of firewood, fodder, and Non-Timber Forest Product (NTFP) for their subsistence and supplemental income but are marginal participants in decision making and in the management of these resources. They are significant contributors to the household's economy by earning agricultural wages and off-farm wages, particularly in areas where male labours have moved out. Generally, women have less access to education than men. This limits their ability to participate in alternative income generation initiatives. Cultural restrictions further impede full and active participation by women in local economies and civic institutions.

But through the Government order during formation of EDCs/ FPCs all the above said vulnerable sections of the PA are empowered to have their representatives and responsibilities in the EDGs/FPCs so that they become active and effective partners of this eco-development project. Composition of the EDC in each village comprises all eligible households including those headed by women, all SCs and STs Women have equal rights with male members. Selected representatives of women, STs/SCs must have been included in the Executive Committee, which maintains a liaison of the registering members with the forest department.

We observed that women are playing more responsible role to protect the forest and ensure success of the programme. From the very psychological phenomenon, as women are creators, they do not favour any kind of destruction. Women, once used to stay within four walls, are now going for regular patrolling with male members to the dense forests to protect woods, animals and prevent poachers.

Individual and Community Project

During making the micro plan, different community development projects such as digging pond and pisiculture, making village roads, embankment of rivers, making *jumpai* (irrigation channels) were considered according to the felt needs of the specific site.

Besides community-based project, several individual schemes have also been adopted. This includes pump machine, sewing machine, grocery, pigery, gotary, poultry etc. All these schemes were selected in that way so that they become eco-friendly and sustainable. The reason behind adaptation of these schemes is to generate alternative income activity so that the dependency on forest will reduce, which will, in turn, increase the green in forest and will help to preserve the biodiversity in future.

GO-NGO Collaborations

NGOs have a critical role to play in the microplanning and implementation process. Frequently, NGOs have entered the arena of PA management through activities to alleviate poverty and improve livelihoods in and around PAs. Their experiences and approaches are necessary addition to the efforts of PA authorities. Furthermore, as non-government institutions, these organizations have their own mechanisms of monitoring and evaluation, which could check the irregularities, and misappropriations of the project activities.

In BTR, one local NGO and one National level NGO have been involved during the preparation of microplanning along with Forest Department. The village members, NGOs and Government authorities were working hand in hand to make the project a success.

Integration of Line Departments

Different Government departments also co-ordinated with the Forest Department, so that poor people could also enjoy other development schemes along with eco-development projects.

From the above discussion, it is now clear that the eco-development project entails a meaningful people's participation of all genders and sections. People have shown their interest to work for the forest eco-system and want to make a balance between man and nature. As a result, the harmful activities for forest like grazing, firewood collection, illegal timber felling, pole theft etc. had been reduced and forest was recovering its green. Through this type of group formation (FPCs/EDCs) an improvement of other social indicators could be noticed in almost all villages in the area. The mobility of women and their literacy rate have increased, different anti-social activities have been abolished, and they are now leading happy and prosperous lives.

III

Problems and Paradigm

Though, in general, majority of the eco-development projects are working nicely, a very good number is also suffering from some problems. The limitations are found among participating FPCs. Some factors that may influence effectiveness of FPCs are: (i) ethnic composition of the village; (ii) economic condition of the members and villagers; (iii) political affiliation of the villagers; (iv) attitude of the panchayat regarding the scheme; (v) literacy level of the villagers; (vi) distance from beat office/range office; (vi) the area under protection; and (vii) type of forest products and their level of productivity.

Some Problems Faced by FPCs

- Lack of confidence and uncertainty about the scheme. This often leads to half-hearted participation.
- Pressure upon lands eventually increases day by day which ultimately hampers the objectives of eco-development project.
- In some FPCs, because of non-availability of alternative sources of employment, people became so dependent on forest products that it became difficult to seek their cooperation in the preservation of forest products.
- Often adjacent villagers (non-FPC members) encroach forest areas even in PAs to collect fuel and fodder to meet their basic needs.

- For many FPCs, alternative grazing lands for the village cattle are not available. In such cases villagers often send their animals to graze in protected forest areas.
- Total dependence on forest resources by some fringe dwellers sometimes leads to generate the frictional unemployment. The dwellers usually suffer when forest cutting has been closed and grazing is not allowed for regeneration of plants or at the time of pruning trees. The Department does not have the financial resources to compensate all such families for their lost income. There is some apprehension that these families, out of frustration and desperation, continue to exploit the forest in an uncontrolled manner.
- Many FPCs feel insecured when members get entangled in some conflicts with poachers/offenders. This is a problem particularly in border areas, where people come from distant areas to cut tress. Some women members have been found to have been lured by the poachers also for clandestine activities.

To cope with these growing problems it is essential to develop among local people an interest about forest and nature conservation to the extent that these people not only accept the idea of conservation but also make it as their own goal. In this regard, participation and self-help are pivotal to striking a balance between forest and nature conservation. There is a need to develop new management strategies, which integrate local people and other stakeholders of relevance to the protected area. This is particularly true about the inhabited transition zones.

Consequently, in more recent measures the aim has been to help to reconcile conservation and utilization interests in the transition zones of forest and nature conservation areas. The LISTRA (Livelihood Systems and Tropical Forest Areas) sectoral project implemented by the Deutsche Gesellschaft fur technische Zusammenarbeit (GTZ-German technical cooperation) has developed a methodology for project activities on that basis. The basic assumption is that the designation of protected areas drastically modifies the livelihood opportunities of the population in the transition zone, a population that is far from being able to respond adequately. LISTRA's concept, therefore, concentrates on strengthening the capacities of rural populations around protected areas. The term *livelihood system* is used to describe the specific social, economic and cultural system within which social groups organize their lives and livelihoods.

People living in transition zones are generally interested in conserving their forest resources, particularly where the forest represents their economic basis. Forest degradation cannot, therefore, be their aim. In fact, it is usually other external influences (e.g. increased settlement density through immigration) which may lead to unsustainable use of the forest. Forest conservation (e.g. by means of restricted use) often simply accelerates the vicious circle of reductions in the resource base, over-exploitation and degradation. The capacity of local people to adjust their actions to such drastic and rapid changes is often not sufficient to cope with the requisite modifications in resource usage, farm organization and village-level social organization. By strengthening their capacities, potentials may be developed and their negotiating positions *vis-à-vis* these of the representatives of nature conservation authorities and other actors improved.

At the heart of the LISTRA concept for building the capacity of people living in transition and buffer zones, five intermeshed methodological elements combine to form a systemic intervention which aims to introduce collective learning processes:

1. *Participatory Innovation Development:* This involves a joint effort between rural people and support services to develop and disseminate technical, social and economic innovations. These innovations can be resource-conserving techniques, new social rules for dealing with certain problems, new ways of creating income etc.
2. *Dealing with Conflicts:* This involves analyzing and negotiation interests among parties, notably among groups living in transition zones and institutional actors such as the nature and forest conservation authority.
3. *Resource-use Planning:* Aims at developing agreements and rules for the sustainable use of natural resources in the areas surrounding forest conservation areas. Such agreements encompass land-use rights and land classification among participants.
4. *Village-level Organizational Development:* Serves to clarify

responsibilities within village-level organizations, in order that agreements and rules can actually be implemented. It is crucial in this context to improve the accountability of local leaders, thus strengthening the representation of village communities within and vis-à-vis the nature conservation authority.

5. *Resource Conservation Extension:* Aims at raising awareness of ecological problems and the impact of certain land-use practices on natural resources.

These five approaches are elements of a systemic *net* illustrated in the paradigm. None of the five *nodes* of the net can be moved without changing the positions of the other four. In the broader context, even this five-node net will not be able to perform its functions effectively without complementary measures such as infrastructure development, institution building, and changes to the political and legal framework enhanced environmental awareness and support for private-sector service providers. Finally, the whole net is held in place by approaches fundamental to the wider project of sustainable development.

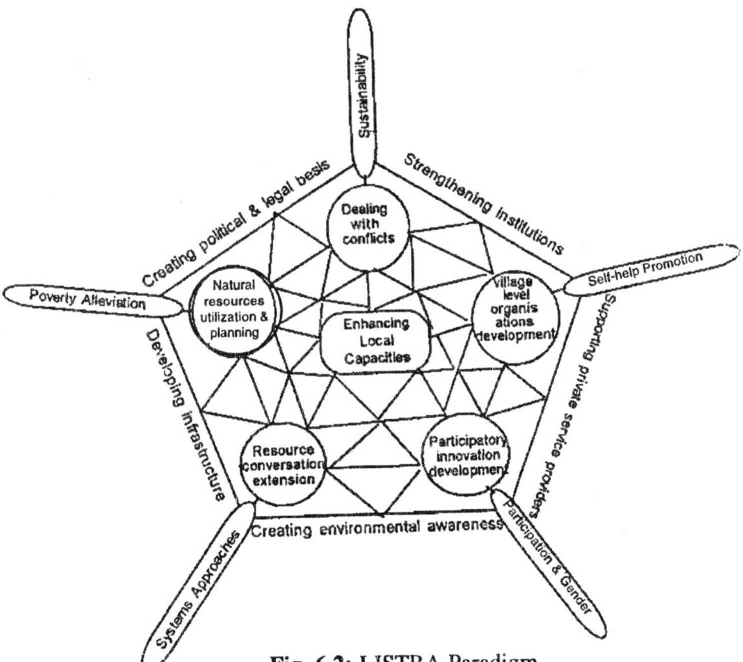

Fig. 6.2: LISTRA Paradigm

REFERENCES

Yugandhar, B.N., and Amitava Mukherjee (Ed.) (1991): *Reading in Decentralized Planning.* (Vol. I & II), Concept Publishing Company, New Delhi.

Rehmen, Kanta, and M.M. Rehman (1998): *Participation for Development,* Manak Publication Pvt. Ltd. Delhi.

Futehally, Laeeq (1988): *Our Environment,* National Book Trust India.

Malhotra, K.C. and Mark Poffenberger (Ed.) (1989): 'Forest Regeneration Through Community Protection: The West Bengal Experience' (*Proceedings*) Forest Department, West Bengal.

Serageldin, Ismail and J. Martin-Brown (Ed.) (1995): 'Servicing Innovative Financing of Environmentally Sustainable Development', Environmentally Sustainable Development *Proceedings Series No. 11,* The World Bank, Washington, D.C.

Rietbergen—Mecracken, Jenifer, (Ed.) (1996): "Participation in Practice", World Bank Discussion Paper No. 333, The World Bank, Washington, D.C.

Dubers, K.N., (1990): *Planning and Development of India,* Ashis Publishing House.

Ghosh, Arun, (1988): "Decentralized Planning: West Bengal Experience". *Economic and Political Weekly,* March 25.

Reddy, V. Reddappa, (1999): "People's Participation and Forest Management in India: FEW Emerging Issues", *Asia-Pacific Journal of Rural Developent,* Vol. IX, December No. 2.

Burger, Dictrich, (2000): "Certifying Forest Management: Sustainable Development in Action", *Agriculture and Rural Development 2/2000.*

Hagmann, J. (2000): "Forest and Nature Conservation" and Development: Putting People first", *Agriculture and Rural Development 2/2000*).

7

Decentralization

Learning from Midnapore, (West Bengal), 1980-2000

AJIT NARAYAN BOSE

Introduction

In West Bengal, Midnapore District has played a leading role not only in the independence struggle but also in understanding the nature of decentralization introduced by our colonial ruler. In the post-independence period also it has played a significant role in the sphere of understanding and implementing Panchayati Raj. The people understood very clearly the dual role decentralization could play. In consequence, they sought to develop a methodology to isolate and use the pro-people aspect of the Panchayati Raj to complement political independence with required socio-economic restructuring of the society, as enjoined also by our Constitution.

After referring to popular resistance to Union Board introduced in 1919, this note deals with three phases of the experimentation on "Village planning by villagers" centred on panchayats by the Zilla Parishad, Midnapore, in close collaboration with the Rural Development Centre of the IIT Kharagpur. This has been preceded by some observations on types of decentralization.

Types of decentralization and serving the people

Colonial Union Board

Decentralization of State apparatus is not a new innovation. It was there even in the colonial days. Leaving aside the urban decentralization through municipalities and corporations introduced in the 19th Century, rural decentralization was introduced through 'Union Boards' as early as 1919. That decentralization did not always lead to empowering the people became evident from the reaction of the people of Midnapore in 1919. There was strong agitation in Contai-Tamluk area against the Union Board. The latter was conceived by the local people very correctly as an encroachment of the colonial state into the affairs of the village society. It was clear even then that one type of decentralization, rather than empowering local people, may in practice become a tool for further centralization by enabling the long arm of the centralized state to reach the grass root level.

Gram Swaraj

If Union Board represents one pole in the range of types of decentralization, then "Gram Swaraj" may be said to represent the other pole. Gram Swaraj is composed of economically self-reliant villages, where all the adult villagers may meet as frequently as needed to plan (a) all the items on which village-wise decision making is possible and execute the same on their own complemented by (b) all the other items where decision can be taken only by combining several lower level units. The suggested plans on these items will constitute the mandate on the representatives to be sent to the various tiers of the upper level. If the representatives fail to follow the mandate, they can be called back as soon as the people at the lower level becomes aware of their failure to pursue the mandate and they elect new representatives.

It may, however, be remembered that even these types of decentralization may not empower the people if the society continues to remain, as it has remained in India even now, class-caste-gender divided leading to the social and economic domination of a tiny upper caste, upper class, male segment over the society. Gram Swaraj in such a society may mean further strengthening of this tiny stratum by making them also politically dominant. This is, in fact, what seems to have happened in India, even with extremely limited "Gram Swaraj" implemented in India through Panchayati Raj Institutions after independence. Even the 73rd and 74th Amendments to the Constitution—making Panchayats and Municipalities the 3rd tier of Government after the centre and states—have not in most cases altered the situation. What then are the alternatives before the people to really empower themselves? Two alternatives have large number of advocates. These are (i) Trusteeship of the socio-economic elite, and (ii) forcible capture of state power by a pro-people party.

Trusteeship of the elite

It is a well established fact that there had been several outstanding individuals in history who gave up their highly privileged socio-economic position to be with the people. These are genuine cases of change of heart. But the trusteeship principle is based upon the faith that the whole—or at least the major part of—the privileged stratum (class/caste/gender) may undergo a change of heart such that instead of exploiting and oppressing the people they get transformed into their trustees. If such change of heart is possible, then, among other things, the massive concentration of property in their hands will be placed at the service of the people, such that at least the economic status of all in the village [or in the urban mahallah] will be similar and all of them can discuss the affairs of their village as equals with the added advantages that in the most difficult initial period of starting the Gram Swaraj the semi-literate toiling majority of the people will have at their disposal all the knowledge and expertise of their elite trustees. If such a thing can actually materialize, we can have a really class egalitarian society without any caste or gender discrimination in an ideal non-violent way.

Forcible captures of State Power by a pro-poor party

This alternative is advocated by those who do not have any faith in the possibility of change of heart of the privileged class in our society.

The central purpose of this alternative is establishing hegemony over the society of a highly centralized party having strong faith in changing the present socio-economic structure. Main task of the party is to capture the centralized state power and use it to restructure the society with the support of the people. In theory, this party is to be primarily composed of the dedicated cadres from among the workers themselves who have been steeped in the struggle of the workers against class –caste-gender exploitation and oppression. In actual practice, however, it has been observed that the party – particularly its leadership, not only in the middle and in the upper tier but also often at the local levels—gets composed of by, hopefully, intellectually "de-classed" people from the middle and other propertied classes. In this system, party leaders control not only the centralized state apparatus but also the other organisations of the working people ranging from trade unions, kishan sabhas, co-operatives etc. to local village councils or municipal wards etc. Such control is said to be necessary to co-ordinate and to orient the specific purposes of organizations of the working people to ensure the convergence of their diverse activities to achieve the central purpose as conceived by the party. This necessity makes the party the highest form of various class of organizations. The initial hegemonic role of party, and then of the state controlled by it, is said to be all the more necessary because of the utter backwardness of the working people. Again, in the initial period, i.e., before the capture of the state power, the centralization of control by the party is conceived as an absolute necessity to meet the centralized challenge of the state apparatus then under control of the privileged classes. Even when the national state comes under the hegemony of the party, this centralization is considered to be necessary to defeat the centralized forces of international reaction.

Despite many vital differences between the above two programmes for serving the working people, there appears to be a family resemblance between the concept of "Trusteeship" and "De-classing" in their crucial reliance on the benevolence of the elite towards the exploited people. Even in this resemblance, there is a sharp distinction between the benevolence of "Trustees" arising from change of heart and that of "De-classing" arising from intellectual conviction. The main antagonism between these two programmes is centred on the role of the centralized party and the state. While the latter constitutes the key elements in the second programme, the first seems to have strong disbelief on the need for reliance either on such a party or a state. In fact, the latter seems to visualize very different type of state, which may not look like the present state at all.

Village planning by Villagers in 1980-85

It was with this implicit background that the experimentation on "Village planning by Villagers" was initiated in 1980 as an internal research project of the 'Rural Development Centre of IIT, Kharagpur in 8 blocks (Panchayat Samities) of Midnapore in close collaboration with the Zilla Parishad.

In 1977, Left Front ministry was firmly installed in West Bengal. One of first tasks was to hold election to the Panchayati Raj Institution after more than a decade. Leftists captured all the Zilla Parishad and most of the Panchayat Samities and Gram panchayats.

After two years of experimentation by IIT-Zilla Parishad the result was placed before a seminar in the later part of 1982 where the Sabhadhipaties (or their representatives) of all Zilla Parishads of West Bengal along with some senior officials and the relevant faculty members of IIT, Kharagpur, participated. They arrived at a unanimous understanding on the basis of facts generated by research that village people can prepare plans of their own village and can monitor its implementation.

By far the most important lesson of this phase of Midnapore Village Planning movements has been the absolute lack of response, despite the above unanimous understandings, by any Zilla Parishad outside Midnapore, to implement this programme of village planning by villagers. It appears that it is not possible for either the potential "Trustees" or the existing party leadership to be enthusiastic about enabling the villagers to prepare their own plan. It seems that the same hesitation has been shown by the left ministry as a whole in not taking a decision on decentralized planning up to 1985, and then when after 8 years of their rule they introduced decentralization, that was only up to the block level where villagers could directly participate in the planning process.

Among other lessons one may mention the following:

Of the two alternatives, viz. (*a*) sending some trained investigators from outside and (b) training local educated boys to collect data from each household, the second alternative has been consciously adopted from the very outset. However, after few months, a crucial lesson has been learnt that training local educated boys means empowering primarily the elite whose children usually get scope for higher education. This has led to two new insights. Firstly, data need to be collected from group meeting and the same need to be analysed involving the group such that the level of organized knowledge for the group can be heightened. Secondly, elaborate data collection need to be replaced as far as possible by group discussion centred on analysing their experiences generating and also identifying the relevant data.

Potential knowledge level of the villagers about their own village is mostly at a much higher level than that of the outside experts or leaders. Most of the latter having a vested interest to sustain their unquestioned leadership over the people have strong inhibition to enable the people to be aware of their potentiality.

There is existence of substantive idle local resources including labour, arising from current mode of production centred on maximisation of private profit.

Initial spontaneous opinions, even of the poorer strata, often reflect not their own interest but the interest of the elite. Thus, there is an objective need for a 'vanguard' to provide some guidance to start with. This very important finding pinpoints the need for a different type of vanguard who will never want to use his superiority in certain aspects to establish leadership over the people blocking thereby 'village planning by the villagers'.

Usual practice of 'people's participation in the planning process' is the participation of people in the plan preparation as conceived by the outside elite. This needs to be turned over. This practice can teach the elite to develop respect for the wisdom of semi-literate rural poor.

Village Based District Planning Process, 1985-87

In 1985, West Bengal State Government took the decision to decentralize State planning upto the Block Level. At that point, Zilla Parishad, Midnapore, took the crucial complementary decision, based on their experience of previous four years regarding village planning by villagers, to prepare their `block plans on the basis of village plans to be prepared by the villagers. In support of this decision, they correctly argued that people could participate directly in the planning or any decision making and implementing process only at the average village level. The crucial need for such participation rose from the joint understanding that (*i*) the existing class-caste-gender-divided social structure is not only highly inequitable but also extremely inefficient in the matter of resource utilization, and (*ii*) this social structure can be turned over to achieve the desired egalitarian society only if there can arise a conscious movement of the rural poor, constituting the bulk of the people, in an organized fashion to achieve the same.

The Zilla Parishad, Midnapore in close collaboration with the Rural development Centre of IIT, Kharagpur, clearly understood that their above decision did not mean a mere quantitative extension of planning activity from their 54 blocks to 10,480 inhabited villages (mouzas) of Midnapore district. It represented a basic qualitative departure from the existing planning process. As a result, they documented their planning methodology in an official document on behalf of the Zilla Parishad and the newly formed District Planning Committee. This document published first in Bengali language in September, 1985, was entitled *Outline of Methodology of Village-based District Planning Process.*

Considering the crucial importance of the above document—which has been updated and published in October, 2000 to constitute the new official guideline for 'Gram Sansad' based Panchayat planning in rural West Bengal—it is necessary to mention the specificities which distinguish this alternative from the planing process now in vogue.

The central core of the usual Planning Process is the action by Government officials and elitist experts with some formal discussion with people's representatives'. In contract, the alternative has the central core in the activity of the villagers themselves with the realization that representative is no equal to the "people" who elect the former, particularly in our type of the class—caste—gender—divided society. That is why, the new document, utilizing the experience of the last few years, identified the '*Gram Sansad*' (It was "village" in the initial document of 1985) as the basic unit of planning emphasising the duality, particularly of this tier of the panchayati Raj organization. The Panchayat, and so the Gram Sansad, is an organ of the government and so the main job in this aspect of their organization is to execute government order coming from the upper tiers. Organizationally stronger and self-reliant system of Gram Sansad seems to have the potential to solve the above inherent contradictions by becoming the basis of an alternative type of state. This potential becomes further strengthened by which 'may not look like the current state at all.'

The process adopted for alternative planning further strengthens this potential. The crucial facts of this process are:

1. To enable the 'Gram Sansad' (GS) to become conscious of the existing production relation as manifested concretely in the class-caste-occupational composition of the area, and the consequent pattern of division of income generated in the area between the working people and the property owning employers and renters.
2. To enable the Gram Sansad to realise how the current system of profit maximising mode of production results in extreme

social inefficiency in managing the productive forces as manifested in huge idle resources as well as in anti-social use of resources.

3. To realize how a far better reality is possible to be achieved in the alternative society with land and other means of production owned, managed and operated by the villagers and other working people themselves producing for satisfying the real needs of the people.

The contrast between 'what' is possible 'and the maximum of' what can be done now' may operate as a strong force to motivate the people to strive for the alternative society for actually implementing 'what is possible'.

This concept of starting with 'what is possible' may appear strange in the context of the current planning process with exclusive attention on 'what can be done now' in this society. But it needs to be remembered that this is precisely the methodology adopted by the National Planning Committee under the chairmanship of Pandit Nehru, in the colonial days. Then the contrast was between 'what is possible' in future independent India and 'what can be done now' under colonial administration. Even in our Constitution, framed by the elite but with a very deep imprint of the recent national and social liberation movement, one can find the attempt to utilize whatever good can be done immediately in a recently liberated colony on the way to reach the alternative socialistic society.

This planning process seems to have presented some aspects of the trajectory to reach 'Gram Swaraj'. As regards "Trusteeship", this seems to have been completely inverted in this process by conceiving the working people—not the elite—as the leader in the fight consciously to save the whole society. In addition to what we may learn from the proposed methodology of planning, mentioned above, there are some vital lessons arising from the attempt to operationalise the above methodology. This includes the following:

The extent of silence of elitism as manifested in the disbelief in the ability of the rural poor for self-reliant activities appears to be very deep among leaders of all political parties, whether in power or striving to be in power. This disbelief may be the basis of their bad faith that unless they remain perpetually in power the interests of the poor can not be served. As a result, the initial lack of enthusiasm of most of the leaders in this 'village-people based planning process' quickly got developed into deep hostility leading them to surreptitiously stop the process barely within two years from its initiation.

Three events seem to have led to the finalization of the above countermove. The first one is the decision of the Zilla Parishad to provide unconditional untied fund of Rs. 50,000 to Rs. 1,00,000 to each of the first batch of nine villages which have been able to prepare their own village plans. This was by the end of 1986.

It may be mentioned that despite central pressure and also the recommendation of their own State Finance Commission in 1996 to give to the Panchayats – constituting the third tier of the government after the centre and the state in consequence of the 73rd amendment to the Constitution in 1992 – untied fund amounting to only 16 per cent of the net proceeds of all state taxes collected and its formal acceptance by the State Government, no such fund has been given to the Panchayat up to December 2000. Such allocation would have meant in 1999-2000 about Rs. 890 crores which come to around, on the average, 2.34 lakhs per village. Hold over finance being the main lever of power, the continuing extent of centralization in the State Government may be clear from the above event. Equally instructive is the visionary boldness of Midnapore Zilla Parishad to allocate untied funds to villages as early as 1986, i.e., long before the 73rd Amendment.

The second event is the publication of a book 'Rural people in Midnapore, 1986' – containing the data generated by the village planning process in Midnapore during 1985-86 and their analysis. The book was jointly published by IIT, Kharagpur, and Midnapore Zilla Praishad. District planning Committee of Midnapore, and Midnapore Planning and Development Society, a voluntary organisation sponsored by the other three. The book had an illuminating forward by Benoy Krishna Chowdhury, the Minister of Panchayat and Rural Development and Land and Land Reforms Department. In this forward the Minister concluded with the following words:

> "...... this report has described in a clear fashion step by step how to prepare a village plan; if Panchayat members and other workers in all the districts carefully study this report, then this village based planning movement can spread with a firm stride in whole of West Bengal".

Startling facts revealed by this report constituted the third event. It included the finding that 45 per cent of the families in surveyed villages were primarily wage workers and if all the government schemes for eradicating poverty including vested land distribution could be properly and fully implemented, the per capita annual income of all these poor families could increase by Rs. 43; but if these workers could have their own self-reliant organization capable of winning the fight to implement the minimum wages act, then this per capita income could increase by Rs. 164; and if this self-reliant organization could spread to the whole of the state mobilizing support of the other working people and other democratic people for enforcing thoroughgoing land reform enabling the villagers to own all the land they till with their own labour, then – even if production remained at the same level, their income could increase by Rs. 226.

Among many others, the above three facts, viz. (*i*) initiation by the Midnapore Zilla Parishad unilateral transfer of untied fund directly to the villages who have prepared their plans; (*ii*) realization by the villagers in the process of formulating the type of village plan advocated by the Zilla Parishad that the most effective way to eradicate their poverty is to build their own self-reliant organisation rather than depending solely on the benevolence of the Government; and (*iii*) clear indication of the Panchayat Minister that the village planning movement will be spread to all the districts of West Bengal—seems to have alerted those in power having the firm belief that state power in their hand is the best mechanism for helping the poor. Hence came the counter attack defeating, it seems, the as yet weak proponents of "village planning by villagers". By 1988, the movement has been quietly stopped from above without any protest by villagers as yet without any effective self-reliant organization of their own.

Despite the overall balance within the stated power structure still being in favour of the centrists, there exists the possibility of making significant local advance if properly motivated minority of political leaders of ruling parties can have the determination to pursue the objective of proper decentralization. They may utilise the formal support which has to be given by state power to decentralization as such because of the 73rd Amendment and also the advocacy of World Bank and several U.N. organizations like UNICEF for grass root planning.

Like Congress ministry in some provinces in colonial India just before the 2nd world war, and left ministry in West Bengal and two other states, it is possible for a country in transition like India (from colonial to independence, from semi-capitalist to beyond capital) to have unusual ministries not committed to status quo. This makes it possible for such ministries to base their planning on a theoretical demonstration of "what is possible" in the desired alternative society and to use it as a foil to contrast the relative meagreness of 'what can be done now' within the present society. Only such a planning methodology can use the planning process for consciencitising the rural people to develop an organized move to have the alternative society. Despite this possibility and necessity, the forces of status quo further strengthened by the various compulsions – including even popular pressure – to concentrate on solving immediate problems and also the lure to remain in power, seem to induce even the initial unusual Ministries to eventually behave as a traditional one for sustaining *status quo*. This seem to compel them to concentrate solely on 'what can be done now'.

What is true about ministries, i.e. political parties in power, is at least equally true about the government officials. Thus, there do exist a small minority of both political leaders and government officials who may have real commitment to move beyond the *status quo*. These contradictions within the government need to be borne in mind to visualize that the proposed planning process may always have to remain in a flux.

Planning Implementation and Aurai Gram Panchayat — Plan–1990-2000

In 1986-87, Zilla Parishad, Midnapore, gave untied fund to nine villages to implement the plan prepared by them in 1985-86 as a part of the district-wise planning movement. This was just few months before the whole planning movement was halted in practice. It seems to have happened through a decision in the major ruling party. As a result, only one village in Contai under the influence of a smaller left could continue with the plan implementation although as an isolated event. This village is Beliachatta.

Centred on the implementation of the village plan for Beliachatta, the local Aurai Gram Panchayat could be motivated to take upon themselves the preparation of 'village plan by villagers' and by 1990 they produced a plan for their Gram Panchayat covering the period 1990-2000. Both the Gram Panchayat plan preparation and its implementation were facilitated by the election of the leading cadre of the Beliachatta planning activity as the *Pradhan* of this Gram Panchayat in 1988. He continued in this position till 1988. A detailed review of this plan and its implementation by the Panchayat and Rural Development Department, Government of West Bengal is now under progress. Pending the finalization of this review, we may mention here few highlights. These include:

Preparation of this plan involved people and, as a result, all the political parties and local people's organizations became integral part of the planning process. So when the plan document was finalised the representatives of all the parties and other organizations of the area put their signature approving the plan document considering it as their own. Co-operation of all political parties in the area ranging from Trinamul to CPI(M) for local development still continues.

Through detailed and concrete calculation the plan demonstrated sector by sector what is possible if all the local resources including labour could be fully utilized to satisfy the people's need.

The plan demonstrated that without any net investment from outside – public or private – the income of the area, at 1988 constant price, could have increased from Rs. 172 lakh in 1988 to Rs. 371 lakhs in 2000. This would have meant an increase of per capita income from Rs. 1262 to Rs. 2276. (P.C. income at 1988 price was Rs. 1680). As the pre-condition for implementing these possibilities is thoroughgoing land reform, the per capital income of the earlier wage worker families would have increased from as low as Rs. 717 to around Rs. 2200 in the visualized alternative society having fully implemented land reform.

Against this concrete foil of 'what is possible' in ten years, if the working people can change the present socio-economic structure, it was demonstrated 'what can be done now'. The above 10 years mean 10 years after full land reform. The crux of 'what is possible' is the full utilization of idle labour as joyous self-employed labour owning their own land and implements. In contrast, 'what can be done now' in the current society depends upon getting government and other outside capital to employ the idle labour as wage labour. Visualizing that compared to around 7 per cent of the plan fund received by the village as tied fund in the immediate past, if the villages would have received its double or 14 per cent of the plan fund as untied fund, this plan formulated 'what can be done now'. The word 'now' here means 10 years, 1990-2000. It was found that even if this higher and untied allocation become possible, only 71 per cent of the idle labour could be employed although only as wage labour certainly with far less productivity than the self-employed labour working on or for their own land. Thus, it would mean far less increase in income than what is possible in the alternative. But this is not its major drawback. The major drawbacks are:

(*a*) It will keep the people dependent upon State making it difficult for the working people to develop their own self-reliant organisation,

(*b*) It will force the state to remain dependent on foreign capital to finance the plan with disastrous consequence for the country as we are experiencing now, and

(*c*) Income increase will be concentrated in the hands of a few at the top and there may not be substantial increase in the wage income, thus condemning them to remain below poverty level.

As regards 'what can be done now', the Aurai GP could implement most of the projects identified by the Aurai plan, 1900-2000 and something more. The existence of a number of schemes identified in the plan, prior involvement and initiative of the people to formulate those schemes, close co-operation of all the political parties in the area in implementing the schemes they themselves have approved earlier – all contributed to this success. As a result, this GP could use all the fund allocated to it by the Government.

This success to implement 'what can be done now' needs to be seen in the context of the failure to implement or move towards implementing 'what is possible'. The cause of this failure is to be firmly located certainly not in the GP but in the failure of the state authorities to implement the programme of "Village Plan By Village People" and to involve the whole state to move towards restructuring their own society which is the primary pre-condition for implementing this aspect of the plan.

This failure seems to have led to a completely unintended and dangerous impact in the area. The Aurai plan was formulated before the adoption of the New Economic Policy. So the Aurai plan could not contain any warning against the dangers that would flow from NEP; and it was not known that to camouflage these damages, the NEP might give some transient opportunity to a few to benefit from eco-degrading and hence unsustainable production of export goods by curtailing production of essential local necessities. Such an opportunity was available to the area in the form of prawn (Bagda) cultivation and export. First to grasp this opportunity was precisely the leading cadres of the village planning movement. And so it could be seen in the area that agriculture land was being transformed into saline water ponds and as yet all – mostly poor farmers who followed the leading cadres, had been making substantive income from Bagda cultivation. The atmosphere in the area now for the last 2/3 years became like one of "gold rush". Main talk in the area had become "how to get rich quickly". An old time capitalist lure seemed to have pervaded the area. The village planning movement in its inception visualized that there would be, as a result of the planning movement, a resurgence of the social change, of going beyond the capitalist society. Exactly the opposite seems to have happened.

Crucial significance of this plan document lay in the fact that it was perhaps the only document at least in 1990 to have concretely contrasted 'what is possible' in the alternative society centred on the hegemony of labour from 'what can be done now' in the current society centred on the hegemony of capital. This document has clearly demonstrated that if the domination of capital can be turned over by labour, then elimination of poverty in a completely self-reliant way, i.e. without any intrusion of foreign capital, may become a certainty. Situation is just the opposite now. Such a poser of two opposites in one frame can become a powerful tool to replace this capital–based society by a labour–based one.

Concluding Remarks

The tentative conclusion from the Midnapore experience (1980-2000), particularly the last phase, appears to be the following:

(*i*) Village (or Gram Sansad) centred planning has the potentiality and necessity to mobilize at least the whole working people to strive for 'what can be done now' only as a step towards 'what is possible' in the altered society beyond capital. Such a planning movement can become a potent tool for a complete restructuring of the present society;

(*ii*) Unless this planning movement spreads out at least to the whole state, the relevance of concentrating on "what is possible" can not be operationalized;

(*iii*) When the relevance of "what is possible" is lost, the whole planning movement may be transformed into its opposite by inducing the people in the isolated areas to conceive the present as a permanent entity including them to concentrate on "what can be done now" in the existing society; and

(*iv*) Leading cadres of the initial village planning movement have the potential to become leading cadres of the above transformation because of their better awareness of the planning technique and their capacity to visualize alternatives feasible within the present society.

8

Women Empowerment for Functional Efficiency in Decentralized Planning at Bottom Level

TAPATI DASGUPTA

Women in India had been subordinated for ages and it was only from the early half of the 20th century that the real struggle for the liberation of women has taken a positive turn. The liberal ideas were already in the air due to the infiltration of Western thoughts. But the ideas could take a matured shape only after Independence. The Indian Constitution, which was framed in 1950 has equipped women with much theoretical strength, but in practice, there lies huge disparity between theory and practice. Devolution of power and democracy are the twin messages of Decentralized Planning. Tuned with the philosophy of democracy, the key to today's administrative set-up is to give shape to people's government based on the consent and goodwill of the grass-root people. The motto of this people's government is to get acquainted with the problems and grievances of the rural mass and to suggest remedies for the major maladies in specific areas. At the apex of the Panchayati Raj administration, men-folk, of course, steer the rural affairs. Till the 1980s the women did not have any specific role in the making of people's government at bottom level. The Nairobi Conference of 1985 and the 1988 National Perspective Plan for Women brought about a worldwide sensation in women's camp. The real breakthrough in women's involvement in politics, administration and development came only in recent years with the promulgation of the 73rd Constitutional Amendment of March, 1993. Some of the States, particularly West Bengal, witnessed a revolutionary change in the status of women after promulgation of the Panchayati Raj Act, 1996 following the introduction of the 73rd Amendment. But the burning question of the current years is how far the women have become really empowered and how far they are capable of exercising their power in political platform. This paper deals with these burning issues and cites many case studies from different States, particularly from West Bengal, in order to examine the actual role and functional status of women in Decentralized Planning.

The women played a vital role in shaping the political as well as socio-economic policies in ancient India. India in the Rig Vedic days witnessed a glorious epoch of prowess for women. In the Vedic period, there were two political assemblies, called the '*Sabha*' and the '*Samiti*' and these political bodies had both male and female representatives. The dignity of women was much lowered in the later Vedic period and the periods following it due to emergence of many social evils in society. Women were regarded as mere items in the household and they were totally under male chauvinistic grip. In the British period too, in-spite of the influx of the liberal ideas, the conservative framework of society could not be much remodelled. The 19th century Bengal Renaissance brought about new waves of changes and gender issue received a new dimensions in the writings of Rabindranath Tagore, Bankimchandra Chattopadhyay, Saratchanda Chattopadhyay and others. Through the Bramho Movement, efforts had been made to re-establish the status of women in a glorious manner. The real breakthrough came in 1917, the year of the Socialist Revolution in Russia, when Sarojini Naidu accompanied by host of ladies went to the British Darbar to place their rights of equal status with men. By the 1921 Act, the housewives who were possessing education and property earned the right to vote. In 1930, an appeal was made for organizing a women's committee for adult franchise. In 1931, at the Karachi Session of the Congress, a claim was made for female education and female franchise.

After Independence and after the framing of the Constitution in 1950, in the 13th, 14th and 15th paragraphs, equal rights of women had been penned down. The Balwant Rai Mehta Committee (1955) for the first time recognized the status of rural women and tried to establish the ways to increase their income and improve the condition of their children. The question of women's participation in rural development process merited attention in 1974 through the uplift of the status of women in India. It considered the question of establishment of statutory All—Woman Panchayat at the village level with real autonomy. The global concern for women's participation, at least in words, was noticed almost at the same time, that is, in 1975, when the

United Nations declared the decade as the Women Development Decade and adopted some resolutions for it. The decision-making power of women was first obtained by women in the Asok Mehta Committee Report of 1978. The Narasingham Committee of 1981 made another progressive effort by declaring that 5 per cent of the elected seats of the Sarpanch would be reserved for women in each Block of each district. In the international arena, the Nairobi Conference of 1985 called on the participating countries to take steps for women's participation in politics through reservation of 35 per cent seats in all elections. Almost simultaneously, that is in 1988 through the National Perspective Plan for Women, the election rights of women had been asserted in India. The real breakthrough in women's involvement in politics, administration and development came only in recent years with the promulgation of the 73rd Constitutional Amendment of March, 1993. Now it will be in the fitness of the things to know how far these declarations and steps regarding the empowerment of women have yielded results in the case of Indian women. In the following section we are discussing a few cases about the treatment meted out to women representative of people in different panchayat bodies. These facts will give an idea bout what the Acts and Declarations intend to do and what the women actually receive from the society.

As case studies we have considered one or two cases of women's status in the local bodies, just as examples, from some States in India. The study shows that the number of cases of maltreatment of the women representatives far outstrips that of rewards or acknowledgement for their activities.

Case Studies

Punjab

Women Panchayat member, Prakash, belonging to the scheduled castes has approached the National Commission for Scheduled Caste & Scheduled Tribe with a complaint of harassment against the village Sarpanch, Kashmira Singh, also belonging to the Scheduled Caste. In her complaint Prakash alleged that on the 8th October, 1999 the Sarpanch demolished her house which was built on land granted by the Gram Panchayat in the early seventies and on the 16th June that year he abused her and got her family members implicated in false cases. While Kashmira Singh has denied the charges, the Director of the Punjab Unit of the Commission has sought the opinion of the Kapurthala SSP on the matter and proof of action taken on the complaint (July, 2000).

Uttar Pradesh

Given the social status of women in Uttar Pradesh, it was heartening to note that more than 50 per cent women candidates made it to the posts of Zilla Parishad Presidents, elections for which were held last month. However, the thrill was short-lived as details about the nature of the candidature of these successful women began to emerge. It was revealed that most of them are wives of ministers, MLAs and political heavyweights of different political parties. Some of them are mothers, sisters, sister-in laws and even girl friends of men in power, who used all sorts of manoeuvering to see them in place. One minister got both his mother and sister elected. Money, muscle and state-power-all were in full play. (August, 2000). A woman ZP member who remains an untouchable, Nirmala Devi, 32, is a Zilla Parishad member in Chandauli district of Uttar Pradesh. She won the last ZP elections with the highest number of votes in the State. Yet she remains an untouchable in the upper strata of society because she was born a Dalit. She explains her lot with rancour, "the victory would have been real if my supporters and co-workers would have accepted the sweets I handed out to them on my winning the elections. They didn't, because I am a Dalit. It made me realize that even after all these years of political activism I continue to carry the 'untouchable' tag. What really hurts me is that people come to me with their problems, seek my help, vote for me, and consider me as an untouchable." (January, 1999).

Madhya Pradesh

A Woman Panch was publicly humiliated. Kunbi Bai, 23 year old Panch of tribal Boregaon village in Betul district, was beaten up and paraded through the streets of her village with her hands tied behind her back and a garland of shoes strung around her neck. Earlier reports said that she was stripped and paraded naked, but Betul Collector contradicted this in a statement issued to the press. Her husband and other relatives, who led the public humiliation, justified the act by saying that it was an old tradition in their tribe to mete out such treatment to a woman who was found indulging in adultery, although

Kunbi Bai protested that she was not guilty and in fact, she had been raped. Eight persons including her husband and brother have been arrested in connection with the incident (June, 2000).

Rajasthan

Woman Sarpanch and four others were injured in a criminal attack. Darshana Rani, Sarpanch Nauganwa in Alwar district, her husband and two other family members were abused and grievously injured on the 1st July in an attack by about 12 persons led by one of her rivals in panchayat elections held a few months ago. The miscreants were armed with guns and iron bars. When neighbour Govindsingh Gurjar, a ZP member, came up to rescue the family, he was also fired at. Residents of the locality staged a 'dharna' and gheraoed the police station in protest against the incident. (July, 2000).

Orissa

Narayanpur Gram Panchayat in Ganjam district in Orissa, headed by B Rajeswari Rao, has banned child-marriage by passing a resolution. Those who deny the order will have to pay a fine of Rs. 1000 and the amount collected on this account will be used on children's education. Those who attend a child-marriage ceremony or support it any way will not be given any post in the panchayat. (July 2000).

Andhra Pradesh

Fatima Bai, Sarpanch of Kalva village in Kurnol district, Andhra Pradesh, has been selected for the United Nations Development Programme's International Race Against Poverty Award for her outstanding contribution to strengthening the poverty alleviation programme in the State. The award was presented to her by the UN Secretary General on October 17, in New York. (Sept, 1998).

Maharashtra

Women Panches share their experiences. About one thousand elected women representatives from different parts of the state assembled in the tiny hamlet of Saigata in Chandrapur district in the first week of March to look back on the five years of women's participation in

Panchayati Raj and take decisions to strengthn their role in the future. Along with success stories like protecting forest resources and fighting illicit liquor trade, women also narrated tales of their husbands beating them and attending Panchayat meetings on their behalf and receiving threats from local goons. Most of the women representatives complained that the agendas for GP meetings are deliberately not given to them and men smirk and pass remarks if they speak in Panchayat meetings. In some parts of the State, no confidence motions against women sarpanches have become a nuisance. The Saigata Convention passed resolution to approach the government on several issues like giving property rights to all women in the villages and urged women to contest also for unreserved seats to increase their seats in the PRIs. (March, 2000).

Karnataka

In an article titled "Women in Panchayati Raj" which appeared in the *Ford Foundation Bulletin,* in 1995, an interesting case study was narrated. A lady called Shantamma was docile, obedient, and subservient to the wishes of her husband and to the male – dominant society of which she was a member. Her duties were not confined just to her family or domestic situation, but also to the panchayat as a member of the Mandal Panchayat, the local village council. It is interesting to note that Shantamma became a member of the Council, which in most parts of the country almost always comprise the men of the community? The credit goes to the State of Karnataka in South India, which led the way in 1987 bringing about sweeping changes in the programme of self-government and reserving 25 per cent of seats in Panchayat bodies at the village sub-district and district levels for women. This was an essential step in enabling the voices of women to be heard in the ordering of affairs at the grass-root level. The ambitious declaration of Shantamma "I want to be the Prime Minister of India", now needs to be echoed in every hut of rural India.

West Bengal

No work for Women: Shramajibee Mahila Samity is a non-party Organisation of rural working women. The members of the Samity have been struggling for many years to get work under schemes like Jawahar Rozgar Yojana (JRY) and Employment Assurance Scheme

(EAS). Even though there are clear assurances in these employment generation schemes that one third of the person-days created under them should go to women, in practice little work is available to them. Also, most Panchayat Pradhans ask the women to do earth work against which there are social taboos in many areas. The Samity has sent a petition to the Minister of State for Land Reforms, Panchayats and Rural Development for creating proper work conditions for women and giving them their share of work under different schemes, especially those types of works that are socially acceptable to them, but no actions were taken on the petition (March, 2000).

A portrait of Change: "I have never yielded to the unreasonable wishes of my husband or in-laws. In the Panchayat meetings I have said what I feel be said, not to please my husband or others", said Kamala Mahato, the soft-spoken Panchayat Pradhan of Bandoan, Purulia District in West Bengal.

Kamala is one of the 186 women Panchayat Pradhans in a State of 3,222 village panchayats. Women won a whooping 59.42 per cent seats in the last Panchayat elections. Women candidates had also won in 17,105 seats in addition to the 24,895 reserved for them by virtue of the 73rd Amendment. However, very few of them were given the top positions. All 16 District Councils in the State had men as heads. Kamala has been a member of the Mahila Samity for 15 years. She looks like any other housewife in Bandoan, coming from a poor village in the poorest district of the State. Sixty-eight kilometers from Purulia town, Bandoan has a mix of tribal and non-tribal population, most of whom are agricultural labourers.

The installation of the new tube wells-a total of 10, in five years in Bandoan since Kamala was elected in the last panchayat polls are among her major achievements. "the scarcity of water has always been a serious problem," she says "We have dug wells, irrigation wells and tube wells in the area." (May, 1998).

On investigation at field level, the primary data reveal the fact that in some of the Gram Panchayats, the women have already acquired a good position and respect. Mention may be made of the All Woman Panchayat at Kultikiri in the Sankrail Block of Midnapore District. There all 10 members are very much confident of their power and position. On interrogation, Tapati Singh, the Pradhan related the fact that since 1992, the Kultikiri All Woman Panchayat is carrying out its sacred task of serving the rural people without any constraint or opposition. Rebati Das, the Saha–Sabhapati admitted that male co-operation from all other quarters is always enlisted in pursuing the activities. Their educational levels are not much high, but they carry out their responsibilities with conviction and faith.

In Laudaha GP of the same Block, Sikha Singh, the Pradhan is administering with much endeavour and efficiency since 1998. The male partners themselves confessed that they face no inconvenience in working with a lady who is at an apex position.

The same picture is being reflected in Gopali GP of Kharagpur 1 Block, where Panmoni Soren, the Pradhan, has proved herself to be a very successful administrator since 1998. The host of male partners, like Phanibhusan Chakravarty, Rakhal Chandra Soren, Atul Maity and Biren Sarkar are fully in accord with the ideas of the female Pradhan and they share all responsibilities jointly.

The picture however is not much redeeming at Kharigamathani GP of Nayagram Block, where her male colleagues have not marked the Upapradhan, Sumitra Shit, as a very assertive lady. The other members also handle the affairs casually and political leaders mostly dominate them.

The above facts, so far analysed, unfold a clear picture of women's changing role in Panchayati Raj. Unfortunately, all the States have not yet started gearing this stupendous task. The States, which are still lagging behind in this experimental scheme, are Arunchal Pradesh, Sikkim, Meghalaya etc. Among the States that have come to limelight, West Bengal is one of the pioneering States, which has set a revolutionary example. In the 1993 election, 24,799 women were elected in the different tiers of the Panchayat. The West Bengal experience appears to be worth emulating and it is now acceptable that in spite of thousands of constraints, a progressive goal can always be achieved. Maharashtra, Madhya Pradesh, Uttar Pradesh, Rajasthan, Orissa are putting their best effort to place the women's right in the forefront of Panchayati Raj code. Unfortunately, Andhra Pradesh is a glaring example of the conservative framework, where only the moneyed women have access to election rights. Only, Fatima Bai's case is an exception in Andhra Pradesh. Himachal Pradesh is not also showing much progressive trend, since women do not claim to have any right in the day-to-day rural affairs.

Today, there are many *Shantammas* who have their prominent voices at Gram Sabha, Mandal Panchayat and Zilla Parishad meetings in different parts of Karnataka. However, the role that women play in the political life of India is still marginal. In the Lok Sabha, only about

40 out of 54 seats are currently occupied by women. Women hold similarly small percentage of seats in state legislative assemblies. Even at the village level women are often either absent or have a token presence. That women are gradually becoming conscious of their rights became evident from the yeaster-years' happenings in Uttarakhand. Women there had been at the forefront at the Chipko Movement, while their husbands were away in the plains seeking employment. They wanted to know why the Sarpanchs, Gram Panchayat members and the co-operative society bosses had to be men always.

Empowering women is not all the solution we look for to fulfil the goals of development. Decision-making, needs, understanding the problem and the rationales of solution are the prime requirements in women. This calls for awareness, education, skill and capability-building among the women who are in large numbers to decide the course of development-planning implementation, monitoring and evaluation at different levels, more particularly at the grass-roots level. This aspect calls for education and training connected with the mundane maladies and remedies of them. Even in the progressive States like West Bengal, Karnataka etc., thousands of women members of Gram Panchayat and Pachayat Samity are in dire need of such capability building. There are many male Gram Pradhans, male Panchayat Secretary or some male members who have become the de–facto administrators of the Panchayat and dictate the path of development in the way they were doing so long without women members. Women, who are brought to the system, should not look for their emancipation only, but should use their skill for the progress of society. What is necessary today is not election or selection, but injection and insertion of skill and capability in them to plan, implement and manage the programmes of development and welfare.

In certain academic, political and social forums, views are expressed that Panchayati Raj has established its roots in society in varying extent depending on the role of particular political party in the State. This view is not factually tenable, since PR has not progressed and become a dynamic developmental institutions across the States with various cross sections of political philosophies. On the contrary, it remained stagnant stereotyped machinery without any vibrant response from people in States with different political philosophies and government wills. Thus, if women have to become an important partner of development and welfare in society, it is not the political philosophy or the party in power, but the acumen and capability in women, which need to be enhanced and refined – at least the 73rd Constitution Amendment has heralded so.

However, there has been no study on either the social background or the role perception of women. But it cannot be denied that given the class structure of the Panchayat leadership, as evident through a number of studies, a good number of women have come out from the lowest rung of social order in terms of either their caste or class. A series of workshops, training sessions and '*melas*' (local fairs) are now being organised by the Government and non-Government agencies to bring together women from different parts of the State and from different levels of local government. This activity has stimulated discussions, encouraged the sharing of problems and experiences and helped in creating a sense of solidarity and confidence.

The most important things that the women are now in the offing to understand the system and are aware of their innate strength and capabilities. Increasingly, they are participating in making decisions once more like that of the Rig Vedic days. This process is in progress from the village or grass-root level to the State level.

NOTES AND REFERENCES

1. Tapati, Dasgupta and R.N. Chattopadhyay, *Rural Development in India: A Socio-Historic Approach,* New Delhi, 1999.
2. State Panchayat Acts, A Critical Review, Voluntary Action Network India, New Delhi, 1995.
3. *Panchayati raj Update,* Institute of Social Sciences (1998-2000 Serites) New Delhi.
4. Tapati, Dasgupta and R.N. Chattopadhyay, The issue of Conflict in Gender Role in India, *Socialist Perspective,* vol. 27, No. 3-4 December 1999, March 2000, Calcutta.

9

Decentralisation of Power

A Case of Women Empowerment

BANHI CHAKRABORTY

Over the last two decades or so concerted efforts are being made to make the women socially and culturally able citizens. The Government of India had incorporated this task of recognizing the rights and privileges of women as its planned task since the Sixth five-year Plan. Successive plans had included different programmes of social welfare for women with thrust on economic up-gradation. Inspite of different supportive programmes the women's status is yet to see any meaningful change. Although it is observed that only 1/3rd of all households (rural and urban together) are supported by women and 33 per cent to 50 per cent of the income of these households is earned by women and altogether 94 per cent of all women is engaged in unorganised sector, a large proportion of the employed women still continue to heavily depend on the primary sector of the economy (81 per cent) while 8 per cent and 10 per cent belong to secondary and tertiary sectors respectively. However, realization of this fact had forced the authority at the governmental level to make a concerted move towards change. The stage set for the change was also constantly being reinforced by a series of UN declarations starting from Nairobi to Beijing and then to New Delhi Conferences of UN at the last meet of the Inter-Parliamentary Union in 1997. The change was seen with the 73rd and 74th Amendments of the Constitution wherein the provision has been made as per articles 243D and 243T for the reservation of 1/3rd of the total seats at all the three tiers of the local-self Government for women. This possibly brought immense hope for a challenging transformation in the social structure. It has been made obligatory for the States to conduct election of Panchayats and Municipalities at regular intervals conforming to all the Articles as mentioned in the 73rd and 74th Amendments. This actually assigns responsibility to the political leaders in-disguise to bring out the women in the forefront.

Background

The commitments have been fulfilled with the inclusion of women at all the three levels of the Panchayats and Municipalities of the States since 1993. However, the present work is involved only with Panchayat women. The 1998 election of the Panchayat Raj Institutions in India brought at least 31 per cent women out of elected members at the Gram Panchayat level while the same at Panchayat Samity level came to around 20.71 per cent and at Zilla Parishad level it was 31.80%. Except the State of UP and Chandigarh all other States achieved the target of 33% at Gram Panchayat levels while the short-falls as observed in Punjab, UP, and Chandigarh and in the States of Assam and UP are respectively for Zilla Parishad and Panchayat Samity level (Table 9.1). However, this success would have been more glaring if the same could have been achieved at the State and parliamentary levels of the Governments. As per United Nations approach for empowering the women, India had succeeded only in one venture while the other three steps, viz., representation at the State and parliamentary levels, at the top level administrative and management jobs (at present only 2 per cent) and in professional and technical services (21 per cent present figure) are yet to be realized. Broadly speaking, the success of the concerted effort of the Government remains less evident in eradicating the social and economic poverty of the women as expressed by their gradual decline in sex ratio (927), lower work participation rate (22 per cent), lower enrollment ratio (47 per cent), low literacy level (39 per cent) and lower share of earned income (27.4 per cent) (Table 9.2). This situation is compelling enough to make one realize that the social inequity exists even after the initiation of structural reforms for the last decade or so.

Following Gidden's theory if we look back to our history it will be seen that "resources are distributed in accordance with structure of domination which very well conformed with the origin of the power relation in our socio-cultural set up that started with the rise of agriculture and concentration of land and other resources including the women (as the property of the male). Social acceptance was subtly manipulated in the favour of the men whom ultimately both covertly and overtly created all social relations and customs in order to centralise more and more

Table 9.1: Number of Elected Women Members (as on May 31, 1998)

Sl. No.	*States/Uts*	*GP*			*PS*			*ZP*		
		Women	*Total*	*%*	*Women*	*Total*	*%*	*Women*	*Total*	*%*
1.	Andhra Pradesh	78000	230529	33.84	5420	14644	37.01	363	1093	33.21
2.	Arunachal Pradesh*	—	—	—	—	—	—	—	—	
3.	Assam	7458	24860	30.00	745	2486	29.07	—	—	—
4.	Bihar #	—	—	—	—	—	—	—	—	—
5.	Goa	468	1281	36.53	—	—	—	—	—	—
6.	Gujarat	41180	123470	33.35	1274	3814	33.40	254	761	33.38
7.	Haryana	17928	54159	33.10	807	2418	33.37	101	303	33.33
8.	Himachal Pradesh	6013	18258	32.93	558	1661	33.59	84	252	33.33
9.	Jammu & Kashmir#	—	—	—	—	—	—	—	—	—
10.	Karnataka	35305	80627	43.79	1343	3340	40.21	335	919	36.45
11.	Kerala	3883	10270	37.81	563	1547	36.39	104	300	34.67
12.	Madhya Pradesh	156181	474351	32.93	3169	9097	34.84	319	946	33.72
13.	Maharashtra	100182	303545	33.00	1174	3524	33.31	587	1762	33.31
14.	Orissa	28595	81077	35.27	1870	5260	35.55	294	854	34.43
15.	Punjab	23892	78983	30.25	326	2441	13.36	89	274	32.48
16.	Rajasthan	38791	119419	32.48	1740	5257	33.10	331	997	33.20
17.	Sikkim	27	1827	1.48	—	—	—	1	145	0.69
18.	Tamil Nadu	31548	94304	33.45	2295	6499	35.31	225	648	34.72
19.	Tripura	1809	5427	33.33	67	196	34.18	24	70	34.29
20.	Uttar Pradesh	174410	682670	25.55	14002	58165	24.07	648	2551	25.40

(Contd.)

Table 9.1 (*Contd.*)

Sl. No.	*States/Uts*	*GP*			*PS*			*ZP*		
		Women	*Total*	*%*	*Women*	*Total*	*%*	*Women*	*Total*	*%*
21.	West Bengal	22014	62172	35.41	3202	9516	33.65	227	664	34.19
22.	A & N Islands	229	667	34.33	25	—	—	10	30	33.33
23.	Chandigarh	22	109	20.18	2	6	33.33	3	10	30.00
24.	D & N Havali	46	135	34.07	—	—	—	4	16	25.00
25.	Daman & Diu	25	63	39.68	—	—	—	5	15	33.33
26.	Lakshadweep #	—	—	—	—	—	—	—	—	—
27.	Pondichary #	—	—	—	—	—	—	—	—	—
	Total	768582	2449759	33.37	38582	129871	20.71	4030	12671	31.80

Note: * Act has not received President's assent.

Elections have not been held.

power for them. The situations that created gender differences with respect to self-esteem between men and women had ended with two different characteristics of power—these are autonomy and sanction which act invisibly and reciprocally". According to Komter, this invisible power mechanism and its operation and consequences generally escape awareness in all societies and India is no exception. The present endeavour to empower the women in order to bring them at par with the men is one such attempt of reversing this age long male doctrine, which may be termed as decentralization of power.

Table 9.2: Women in India

1.	Percentage share of earned Income Per Women has increased from	25.4% in 1995 27.4% in 1997
2.	Percentage of Adult literacy rate	37.7% in 1995 39.4% in 1997
3.	Life Expectancy at birth	61.8% in 1995 62.9% in 1997
4.	Percentage of Gross enrolment ratio	46.5% in 1995 47.0% in 1997
5.	Percentage of work participation rate	22.7% in 1991
6.	Percentage of female Administrators & managers	2.3% in 1997
7.	Percentage of female professional & 6technical works	25.5% in 1997
8.	Percentage of female in Parliament (both houses)	8.3% in 1997* 8.9% in 1997*
9.	Percentage of women in the Panchayat System	GP—31.37%** PS—20.71%** ZP—31.80%**

Source: World Development Report—1998

Empowerment

An *empowered women* must be educated, informed, skilled, productive, aware of options, and decisive and able to exercise her choice in addition to her existing role. Integration of all these ideas in the process of development for its effective realization is called the process of empowering the women.

An empowered woman is:

- educated and informed,
- skilled,
- productive,
- aware of options,
- able to exercise rights and choices,
- capable of transferring the sense of ethical and moral values to future generation,
- dignity is respected without any prejudices or conditions.

While a powerless woman is:

- illiterate
- unskilled
- physically weak
- with lesser productivity
- dependent
- ignorant of her rights and choices

In the context of the above discussion one might like to know how far the objective of empowerment of women has been achieved under the new dispensation of the Panchayati Raj Institutions (PRIs). In the following paragraphs we have taken up this issue for an empirical analysis with field survey data collected from some PRIs which were selected from different blocks of the District of Midnapore, West Bengal.

Women and PRIs of West Bengal

West Bengal is one of the States in India where women participated in the Panchayat elections before the 53rd Amendment of the Indian

Table 9.3: Percentage of Elected Women in the Last Election, 1998. WB

Districts	*Three-Tier Panchayat*		
	Zilla Parishad	*Panchayat Samity*	*Gram Panchayat*
Bankura	33	35	36
Birbhum	34	33	36
Burdwan	34	34	35
Coach Behar	33	35	35
S. Dinajpur	33	35	35
N. Dinajpur	35	34	36
Hooghly	34	34	36
Howrah	38	33	35
Jalpaiguri	34	35	36
Malda	33	34	38
Midnapore	33	34	36
Murshidabad	33	34	36
Nadia	34	34	35
S. 24-parganas	34	34	35
N. 24-parganas	34	35	35
Purulia	35	34	36
West Bengal	34	34	36

Constitution was enacted. Participation was remarkable in the initial year (1993) ranging from 34 per cent at Panchayat Samity and Zilla Parishad levels to 35 per cent at the Gram Panchayat level. The spirit of the same was maintained even in the next Panchayat election (1998) also (Table 9.3). The strong political base indeed has been instrumental in yielding outstanding result and has probably helped in instilling the confidence in the minds of thousands of women. It is, therefore, expected that women's political emancipation will bring into effect the social and cultural upheaval in the near future not only among the political participants but also among the vast community of common women. Out present study seeks to focus on this aspect of women's empowerment.

Methodology

Data Collection

The study is based on primary data collected through "case study method". This method helps the investigator to get an insight into the actual working of the system in a specific situation. The data were collected by interviewing women functionaries ranging from ordinary members of Gram Panchayat or other Panchayat bodies to Upa-Pradhan and Pradhan at G.P. level, Sabhapati at the Panchayat Samity level. Table 9.4 shows the distribution of women for the two consecutive election years in the District of Midnapore while Table 9.5 shows the same for the selected blocks in the district. Panchayat members were selected from six different blocks, viz. Kharagpur-II, Keshpur, Panskura I, Sutahata 1, Debra I and Sankrail. Frequent visits helped in accumulating information on their working pattern and their experiences.

Table 9.4: Proportion of Women at different tiers of Panchayat in the District of Midnapore.

Year	*Gram Panchayat*		*Panchayat Samiti*		*Zilla Parishad*	
	Total	*Women*	*Total*	*Women*	*Total*	*Women*
1998	7188	2575 (36.00)	1295	448 (35.00)	108	36 (33.33)
1993	9043	3139 (35.31)	1494	503 (34.00)	108	37 (34.26)

Table 9.5: Women Representation in Panchayat Bodies (in study area)

Areas	*Block*	*1993*		*1998*	
		Total	*Women*	*Total*	*Women*
(*a*) Sutahata PS-1	Sutahata PS-1	16	6 (37.5)	15	8 (53.33)
(*b*) Kultikri GP-1	Sankrail PS	11	11 (100.00)	8	8 (100.00)
(*c*) Amrakuchi GP-8	Keshpur PS	17	6 (35.29)	14	5 (35.17)
(*d*) Debra GP-1	Debra PS-1	17	7 (41.8)	13	5 (38.46)
(*e*) Changual GP-1	Kharagpur PS-2	21	7 (33.33)	16	6 (37.50)
(*f*) Panskura GP-1	Panskura PS-1	20	7 (35.00)	16	6 (37.50)

Source: Data collected from the office of the District Planning & Development, Midnapore, 1998.

The data were collected in two parts: (*i*) The first set of questionnaire takes into account the personal status of the respondent, income, family background, historical background of her entry into politics, personal activity status and work adjustment, involvement in political activities and family support, outside exposure of the respondent, interaction with the public, level of fulfilment of her commitment, issues of major achievements and her role at the time of emergent problems (Table 9.6). (*ii*) Details of information at Gram Panchayat level were collected by using the second set of questionnaire. The information on the total population, class, caste, educational levels, social status of the women, distribution of benefits, general problems of the area, expectations of the women folk from the respondents were collected from the Panchayat offices as well as from the common people.

Information was also gathered informally from the officials of the Panchayat in order to get an idea about their perception and feelings regarding the women representatives as well as from the general people for knowing about their performances.

Owing to absence of sufficient data, no statistical techniques could be applied. However, the available data satisfied the purpose of qualitative comparison.

For the purpose of measuring the indices of competency and empowerment, qualitative and quantitative data, wherever available, on the following characteristics were collected:

Table 9.6: General Information about the Respondents

Name of the Respondent	Address or Location with GP Name	Age	Caste	Position Held	Level of Education	Self-Employment Status	Self-Income per month	Employment Status of the family	Total Income per months of family (Rs.)
1. Maya Mukherjee	Dakshin Mechoram Dehati Panskura I	42	G	Pradhan	B.Sc. B.Ed.	School Teacher	4000/-	School Teacher	9000/-
2. Pradhabati Paria	Changual GP No. 1 Kharagpur 2 Kharagpur PS Amrakuchi	45	G	Pradhan	(X)	X	X	Massion	8000/-
3. Chhabi Douli	GP No. 8 Keshpur P Samity Horkhali	50	SC	Ord Member	(VIII)	DWCRA150/-	150/-	Wage Earning Domestic Aid	3000/-
4. Sonali Mallick	Sutahata-1 Panchayat Samiti	32	SC	Sabha	HS	N	N	Agri Land of 10 Bighas	5000/-
5. Rebati Das	Kultikri village Kultikri GP No. 9 Sankrail P Samity	32	G	Pradhan	(IX)	N	N	Tailoring Business Agri Land of 30 deci.	
6. Tapti Singh	Mandar vill. Kultikri GP No. 9 Sankrail Baligeria	27	ST	Upa Pradhan	SF(F)	Cultivation	2000/-	N	2000/-
7. Sandhya Paria	Kultikri GP N. 9 Sankrail Lata Vill	33	G	Ord Member	(VIII)	550/-	Cooking Assistant	1550/-	
8. Repe Mumu	Kultikri GP No. 9 Sankrail Nakodona	25	ST	Ord. Memb.	(X)	N	N	Business of Pen	1000/-

(*Contd.*)

Table 9.6 (*Contd.*)

Name of the Respondent	*Address or Location with GP Name*	*Age*	*Caste*	*Position Held*	*Level of Education*	*Self-Employment Status*	*Self-Income per month*	*Employment Status of the family*	*Total Income per months of family (Rs.)*
9. Dipur Pradhan	Vill. Kultikri GP No. 9 Sankrail	24	G	Ord Memb.	SF	N	N	Cultivator	3000/-
10. Chandana Douli	Vill. Baragath PO Debra bazar Debra I Debra PS	33	SC	Pradhan	(VIII)	N	N	Service	4200/-
11. Dalia Bhattacharya	Vill. Chakpapan P.O. Chak Narasingh Debra Debra PS	27	G	Upa Pradhan	Graduate (B. Com.)	Cultivator as share Cropper	2000/-	Do	Do

1. *For Measuring Competency*

   (*a*) Whether able to visit different offices for official purpose.
   (*b*) Whether able to inspect personally different developmental activities.
   (*c*) Whether able to intervene personally in solving different problems.
   (*d*) Whether able to interact with different organizations.
   (*e*) Whether capable to interact with local people.
   (*f*) Whether non-hesitant to move in areas of trouble.
   (*g*) Whether able to take up challenging tasks.
   (*h*) Whether able to take independent decisions.
   (*i*) Whether able to manage office (Panchayat).

2. *For Measuring Empowerment*

   (I) In *Social Life*

   (*a*) participation in public life.
   (*b*) able to exercise choice for social development programmes.
   (*c*) able to take decisions on public affairs.
   (*d*) able to apply own judgement.
   (*e*) wide acceptance in the society.

   (II) In *Personal Life*

   (*a*) able to exercise choice for family's welfare.
   (*b*) feeling of self-dependence.
   (*c*) able to take decisions in personal matter.
   (*d*) liberal acceptance by the family.
   (*e*) right to parental property/resources.
   (*f*) right to in-law's property/resources.
   (*g*) able to exercise own views with regard to family expenditure plan.
   (*h*) right to spend from own earned money at own discretion.
   (*i*) able to spend for own needs from self-earned income.

Analytical Observation

The following are our findings obtained from the analysis of data and information:

(a) Women in general who had been inducted in the Panchayat Raj system had taken full interest to do the job. Their honesty and sincerity, as had been reported by the staff of the panchayat offices, has brought about a highly congenial atmosphere not only in the offices but outside also, while interacting with the people, which was not in existence earlier. Their sense of accountability to the office is acknowledged by all (based on the interviews of the Office Staff).

(b) Women, whoever came for political participation through electoral process, in fact, were nominated by the local political parties. For eligibility of the candidates, family affiliation with political parties played the major role (Table 9.7). Level of education or knowledge base as one of the major criteria was by-passed on many occasions, even the candidates having lower qualifications were chosen as Pradhan in some areas (vide Table 9.6).

Table 9.7: Factors Contributing to Women's Entry in the Panchayat

Sl. No.	*Factors (% of contribution of indicators)*		
	Family Background	*Self Status*	*External Political Influence*
Office Bearers			
1.	67	100	100
2.	33	50	50
4.	100	12	100
5.	40	100	100
6.	20	100	100
10.	100	62	50
11.	20	100	50
Av.	55	68	79
Ordinary Members			
3.	17	50	50
7.	17	100	100
8.	33	100	100
9.	83	50	100
Av.	38	75	88

(*c*) In certain cases the women with higher perception and education, when allowed to act freely, could exercise their power in day-to-day functioning of the Panchayat and also could mobilize the local political leaders (Table 9.8).

Table 9.8: Organising Skill of the Respondents

Sl. No.	*Health & Sanitation*	*DWCRA Group Form*	*Land Reform*	*Asset Building*	*Spread Literacy*	*TRYSEM trg*	*De–alcoholism*
Married							
1.	1	1	0	1	1	1	1
2.	1	1	0	0	1	0	1
3.	0	1	0	0	0	0	1
4.	1	0	0	1	0	1	0
8.	0	0	1	1	1	0	1
9.	0	0	1	0	1	0	1
10.	1	1	0	0	1	0	1
Av.	(51)	(51)	(11)	(40)	(70)	(11)	(81)
Un-married							
5.	1	1	1	1	1	1	1
6.	1	1	1	1	1	0	1
7.	1	1	1	1	1	0	1
11.	1	1	0	1	1	0	1
Av.	(100)	(100)	(70)	(100)	(100)	(25)	(100)

Notes: Figures in the brackets are the percentages of respondents attributed.

(*d*) The contributions of the officials (Job assistant and Secretary) of the Panchayats to skill formation and capacity building among the women are worth noting. In the absence of these contributions, management of offices by these timid and unskilled women would have never been possible.

(*e*) Another significant aspect as observed among these women members at different levels of village administration is that no one is free from family and domestic responsibilities (Table 9.9) and scales of these responsibilities vary depending on the economic status of the families. However, these socially inherited responsibilities for being women stand on the way of women's participation, which is quite unlikely to happen in the case of male members involved in the political process.

Table 9.9: Distribution of Workload and Involvement in Panchayat Activities

Sl. No.	*Marital Status*	*Position Held*	*Time Spent in Different Activities (% of total time-spent) Panchayat Related*					
			Panchayat Office	*Social Work Excl. Panch.*	*Party Activ.*	*In Job*	*Domestic Activ.*	*Total Working Hours/day*
1.	M	Pradhan	25.00	7.00	10.00	25.00	33.00	16.00 (100)
2.	W	Pradhan	40.00	16.00	27.00	Nil	12.00	12.30 (100)
3.	M	Ord. Mem.	Neg.	Neg.	30.00	26.00	44.00	10.00 (100)
4.	M	Sabhapati	22.00	Neg.	28.00	Nil	50.00	10.30 (100)
5.	UM	Pradhan	40.00	16.00	20.00	16.00	8.00	12.30 (100)
6.	UM	Upa. Pradh.	37.50	12.50	19.00	20.00	12.50	16.00 (100)
7.	UM	Ord. Memb.	11.00	11.00	22.00	58.00	Nil*	9.00 (100)
8.	M	-do-	Neg.	11.00	22.00	Nil	66.00	9.00 (100)
9.	M	-do-	Neg.	Nil	20.00	Nil	80.00	10 (100)
10.	M	Pradhan	33.00	Neg.	11.00	Nil	59.00	13.30 (100)
11.	UM	Upa-Pradh.	43.00	7.00	7.00	28.00	15.00	14.00 (100)
av. of all =			24.00	7.00	18.00	17.00	36.00	100.00
			(3.00)	(1.00)	(2.00)	(3.00)	(3.00)	(12.00)*
av. of married=			19.00	5.00	20.00	9.00	47.00	100.00
			(2.00)	(0.30)	(2.00)	(3.00)	(5.30)	(13.00)*
av. of unmarried=			35.00	12.00	17.00	26.00	11.00	100.00
			(4.30)	(1.30)	(2.00)	(4.30)	(1.00)	(13.30)*

Note: Figures in the brackets (*) indicate the actual time in hours.

Table 9.9*(a)***:** Recognition in the Family

Sl. No.	*Recognised by*		
	Parents	*husband*	*in-laws*
Married			
1.	f	f	f
2.	nil	n.r	n.r
3.	f	n.r	p
4.	f	f	f
8.	f	p	f
9.	f	n	p
10.	f	f	f
Unmarried			
5.	f	n.r	n.r
6.	f	n.r	n.r
7.	f	n.r	n.r
11.	f	n.r	n.r

Note: f=fully, p=partially, n.r=not recognised

Table 9.9*(b)***:** Criteria Behind Recognition

Sl. No.	*Income Contribution (% to total family income)*	*Fulfilment of Household Duties*	*Family's Political Connection*
Married			
1.	44	f (5 hrs)	y
2.	0.0	p (1.5 hrs)	n
3.	0.7	f (4.8 hrs)	y
4.	0.0	f (5.0 hrs)	y
8.	0.0	f (6.0 hrs)	y
9.	0.0	f (8.0 hrs)	y
10.	0.0	f (8.0 hrs)	y
Unmarried			
5.	53	p (1.0 hrs)	y
6.	35	p (2.0 hrs)	y
7.	33	nil*	y
11.	100	p (2.0 hrs)	y

Note: f = fully, p=partially, y=yes, n=no

* The respondent works and stays in a girls hostel for which she is spared from domestic responsibility.

(*f*) Women with sound economic background and higher cultural status and higher educational attainment scored high on the performance scale (Table 9.10a and 9.10b). Women with poor economic status and low level of education, although found to be sincere in their duties, were less assertive in the exercise of their power. Maya Mukherjee, Pradhan of Panskura I GP and Dalia Bhattacharya, Upapradhan of Debra-I GP provide example for the first case while the name of Prabhabati Paria can be cited for the second case.

(*g*) The political parties, however, proved to be little judicious in considering the candidates during 1998 election. In most cases the women who couldn't show much efficiency during the first term because of their commitments towards the families did not get nomination for the second term.

(*h*) It is significant to note that in no case women were allowed to contest against the general seats which were earlier reserved, even with their proven efficiency during their first term.

Table 9.10*(a)*: Competency Level of the Respondents

Sl. No.	*Indicators*										
	1	*2*	*3*	*4*	*5*	*6*	*7*	*8*	*9*	*Composite Scores (100)*	*Rank*
Other Bearers											
1.	1	1	1	1	1	1	1	1	1	100	1
2.	1	1	1	1	1	1	1	0	0	78	2
4.	0	1	0	0	0	1	0	1	1	44	3
5.	1	1	1	1	1	1	1	1	1	100	1
6.	1	1	1	1	1	1	1	1	1	100	1
10.	1	1	1	1	1	0	1	-1	1	78	2
11.	1	1	1	1	1	1	1	1	1	100	1
Ordinary Members											
3.	0	1	1	0	0	1	0	nr.	nr.	43	2
7.	1	1	1	1	1	1	1	nr	nr	100	1
8.	1	1	1	1	1	1	1	nr.	nr.	100	1
9.	0	0	1	0	0	0	0	nr.	nr.	14	3

Note: nr = not required.

Table 9.10(b): Competency Values in relation to Marital Status

Sl. No.	Indicators										
	a	*b*	*c*	*d*	*e*	*f*	*g*	*h*	*i*	*Composite Scores (100)*	*Rank*
Married Members											
1.	1	1	1	1	1	1	1	1	1	100.00	1
2.	1	1	1	1	1	1	1	1	2	78.00	2
3.	0	1	1	0	0	1	0	nr.	nr.*	43.00	4
4.	0	1	0	0	0	1	0	1	1	44.00	3
8.	1	1	1	1	1	1	1	nr.	nr.*	100.00	1
9.	0	0	1	0	0	0	0	nr.	nr.	14.00	5
10.	1	1	1	1	1	0	1	0	1	78.00	2
Unmarried Members											
5.	1	1	1	1	1	1	1	1	1	100.00	1
6.	1	1	1	1	1	1	1	1	1	100.00	1
7.	1	1	1	1	1	1	1	nr.	nr.	100.00	1
11.	1	1	1	1	1	1	1	1	1	100.00	1

Note: Competency values: "1" for yes, "0" for negative, nr. = not required.
Indicators of Competency:

(*a*) Able to visit different offices for official purposes.
(*b*) Able to inspect personnlay different developmental activities.
(*c*) Personnel intervention in solving different problems.
(*d*) Able to indicated different requisition
(*e*) Interaction with local people.
(*f*) Non-hasitation to move in areas of trouble.
(*g*) Able to take up challenging tasks.
(*h*) Able to take important decisions.
(*i*) Able to manage office (Panchayat)

Emerging Trend

Women's Participation

The way women came to political process since the implementation of 73rd Amendment of the Constitution, left little scope for them to be true representative of the people. The very basis of selection made by the political leaders was the conventional rule of class, caste and political affiliation. It is most likely that their decision under such circumstances has to be governed by the party leaders the majority of whom are male. However, in the present form of democracy, there is no other way to derive opportunity for the women where they could express their thinking freely. Moreover, to make them more participative, the necessary arrangements are required to provide in every sphere of the society for which the society is yet to get ready. Probably, this may be the reason why all of them, except one, decided neither to contest in 1998 election, nor to remain involved in active politics.

Sensitisation of Women

Under the present circumstances, it would be equally unreasonable to think that the women who came to political platform are sufficiently sensitised. Prohibition on early marriage, lower educational level, dowry system, excessive domestic workload – all these gender specific issues are yet to be included in the work programme of the Panchayat for strategic decisions. Women who are in power are yet to overcome their age-old inhibition in this regard although it was expected from them with the inclusion of the women in the Panchayati Raj system. However, de-alcoholism in some cases was pursued with success and this appeared as significant landmark in the electoral process as the women did not use this as a weapon for vote banks. This was made possible only because they could substantially bring down alcoholism by shuttering down the shops as well as by mass protesting against their spouses. This activity gave them an edge over the male. However, the women are less successful in the cases of wife—beating and dowries.

Recognition by the Family

In spite of fulfilling of all household duties and obligation women are yet to get recognition totally in their families irrespective of their social ranks. However, the women with earnings could show little command over these families. In fact, several factors determine the acceptance by the families of which marriage is one of the crucial determinant. In the given situation it is difficult to deny the fact that married women had to work under greater pressure of family rules and regulation which was almost absent in the case of the unmarried women (Table 9.11).

Table 9.11: Recognition in the Family

S. No.	*Recognition by*		
	Parents	*Husband*	*In-laws*
Married			
1.	f	f	f
2.	Nil	n.r	n.r
3.	f	n.r	p
4.	f	f	f
8.	f	p	f
9.	f	n	p
10.	f	f	f
Unmarried			
5.	f	n.r.	n.r
6.	f	n.r	n.r
7.	f	n.r	n.r
11.	f	n.r	n.r

Note: f = fully, p = partially n.r = not required.

Fulfilment of Aspiration

In spite of women leaders' low level of perception regarding the gender specific issues, their contribution in bringing a change in the mindset of the common women folk merits special mention. They now can come out to talk with their representative, although socially imposed restriction in the family life still exists. In general, women's representation in the Gram Sansad meetings was found to be insignificant but in the specially conducted meetings the participation was satisfactory.

Acceptance by the Society

Women with no obligation in in–law's house and those who are not married were found to have better response in public matters with, of course, one or two exceptions (Maya Mukherjee of Panskura Panchayat; for example). Society is yet to give the women respite from their workload at home and full freedom to participate in public affairs (Table 9.12). Moreover, the women are to face tough competition with the male opponents in decision making process as male supremacy is prevalent everywhere. There are many instances of Gram Sansads' meeting as well as other meetings at different levels of the Panchayat System and also at the top most

Table 9.12: Status of Empowerment – The Respondents Achieved

	In Social Life							In Personal Life											
S. No.	*a*	*b*	*c*	*d*	*e*	*Total Scores*	*Rank*	*a*	*b*	*c*	*d*	*e*	*f*	*g*	*h*	*i*	*j*	*Total Scores*	*Rank*
1.	1	1	1	1	1	5(100)	1	1	1	0	1	1	0	1	1	0	0	6(60)	5
2.	1	1	0	0	1	3(60)	4	0	0	0	1	0	1	0	0	0	0	2(25)	9
3.	1	1	nr	0	1	3(75)	2	0	0	1	0	0	0	1	0	1	0	3(30)	30
4.	1	0	1	0	1	3(60)	4	0	1	1	1	1	1	0	nr	nr	0	5(62)	4
5.	1	1	1	1	1	5(100)	1	1	0	1	1	1	nr	1	1	1	0	7(77)	3
6.	1	1	1	1	1	5(100)	1	1	1	1	1	1	nr	1	1	1	0	8(88)	1
7.	1	1	nr	nr	1	3(100)	1	0	1	1	1	1	nr	0	1	0	0	5(55)	6
8.	1	0	nr	nr	1	2(66)	3	1	0	0	1	0	1	1	nr	nr	0	4(50)	7
9.	0	0	nr	nr	1	1(33)	5	0	0	0	0	1	0	0	nr	nr	0	1(12)	10
10.	1	1	0	0	1	3(60)	4	1	0	0	1	1	1	1	nr	nr	0	5(62)	4
11.	1	1	1	1	1	5(100)	1	1	1	1	1	1	1	1	1	0	0	8(80)	2
Avg. Score & Rank						69	2.4											49	5.3

Note: Percentages have been calculated excluding the not required (nr) responses. 1 indicates positive response, 0 indicates negative response.

Indicators: *Social Life* – (*a*) Participation in public life, (*b*) able to exercise choice for social development programmes, (*c*) able to take decisions in public affairs, (*d*) able to apply own judgment, (*e*) widely accepted by the society.

Personal Life – (*a*) able to exercise choice for family's welfare, (*b*) feeling of self-dependence, (*c*) able to take decisions in personal matter, (*d*) liberal acceptance by the family, (*e*) right to parental property/resources, (*f*) right to in-laws property/resources, (*g*) able to exercise own views with regard to family expenditure plan, (*h*) right to spend from own earned money at own discretion, (*i*) able to spend own needs from self-earned income, (*j*) able to control own working hours (working not more than 8 hrs.)

Government and bureaucratic levels where men's decision get priority and sanction even today because of insignificant representation of the women. However, women have to face similar constraints even in the open market situation which bears potential threat for the Government – sponsored self-reliance schemes for income generation, as is evidenced from study on DWCRA women. It is interesting to note that any programme concerning the women is least attended by the women and decisions are made by the male Panchayat members.

Empowerment of the Women

If the empowerment is aimed at bringing in high esteem and self-emancipation of the women, the women, at least for the present case, are yet to achieve this goal. The gender-biased training that starts from their early childhood or infancy leaves little scope for them to feel the differences and by the time the realization dawns upon their mind, they are transferred to other's family, which virtually enforce them to surrender to the male folks completely.

It is significant to note that even women with money earning capabilities are neither spared from the state of drudgery nor can they exercise their right to spend from their own earnings. The little respect that they receive from their male counterparts or in-laws is because of their role as economic providers to the families.

This will be a hard reality for one to fully cherish the state of empowerment only through political freedom unless she can exercise her rights in her own family (Table 9.2). The findings also suggest that social set-up is yet to be ready to accept as well as accommodate the changes once the women are empowered. Under this circumstances it will not be wrong to seek clarification regarding the genuineness of the so called social leaders' verbal stunts and to ask for the extent they are prepared to allow for their wives' emancipation in reality. Will it not be a blow to their ego and established images in the society if the women at their home disagree to sacrifice their self esteem any more at the cost of which all leaders so far attained social esteem and raised slogans for equality and justice for the deprived lot? Can one deny the fact that political freedom could hardly bring relief in their personal life? It is true in all respects that unless there is total freedom and if the efforts are not made to restore the lost dignity of the women both at the public and domestic life, the talks of women's emancipation will be on paper only contended with political induction.

Therefore, with regard to the acceptance of the women's entity at per with that of male, the society is yet to overcome a number of inhibitions like the socio-cultural status of the women, economic background of the families, conventional taboos and norms which go against women even within the family and self-realization by the women themselves regarding their importance in the society. The attitude of negligence in which they have been brought up had imbibed in them a sense of inferiority and overcoming this will require conscious effort on the part of the people of both sexes to recognize the importance of the role of women in the economic development of our country.

Nevertheless, the notable facts, which came into prominence because of implementation of programme for "women empowerment" need to be mentioned here. These are:

(*a*) Induction of women in the public affairs has contributed to the formation of a new battalion, more aptly, for political parties than as a critical mass necessary to generate a force in establishing their rights of social justice because of its monetary return.

(*b*) Demand for women domestic chores leaves little option to make them free for political participation or involvement in public affairs.

(*c*) Male's indifference grossly act as a deterrent factor for mass involvement of the women.

Policy Decision

The momentum that gathered through their induction in political system now must be maintained in order to bring a regiment of leaders sufficiently ignited for the cause of the society. Therefore, there is a need for a move towards careful and cautious policy decision and these are:

(*i*) A need for fresh thinking on the reservation policy in order to make the women – a real strong and sustainable force of humankind. Considering their actual strength in the total humanity it is desirable to make fifty per cent instead of present practice of thirty three per cent.

(*ii*) Irrespective of caste and class differences women should organize and unite themselves in one unit such that they get a chance to interact freely and select their representatives without any interference from their male counterparts.

(*iii*) The tendency to withdraw oneself once the tenure in the Panchayat is over, as is commonly seen in our study area, should be stopped. In fact, the first phase of their tenure basically acts as a period of apprenticeship. Therefore, to develop an appropriate mindset one needs more involvement and exercise of power. Under such condition, these women (ex-Panchayat members) should be provided with proper opportunities to form a nucleus of power which further may act as a dissemination centre for other women to make them self-reliant.

(*iv*) These women also may act as interface between the Panchayat and common people and in the process may be able to act as a catalytic agent to rejuvenate the Gram Sansads which are at present almost in dying stage.

(*v*) It is desirable to reduce the domestic workload in order to enhance the efficiency level of the women to that of men. To achieve this, an effort at the Governmental level should be made to provide support services for child-care. At the same time traditional customs and social perception which affect women's aptitude for development must be eliminated. However, this requires a change in a mindset of the male folk, which can be achieved by organizing frequent meetings and counselling. Panchayat's initiative in this respect is an essential requirement.

(*vi*) The political emancipation can not be an independent measure for women development until their separate entity is fully acknowledged and respected at home. To achieve this, an all-out effort must be made not only at the governmental level but also at other levels in the society and further down at the family level. However, unless the women, the victim of social taboo, come forward banishing age-old hesitations and take bold initiative, no Constitutional revisions can help them to emancipate.

Several options rest with the institutions to bring changes in the present social and organizational set-up. These are:

- allowing the time flexibility in the Panchayat offices so that they can adjust with their domestic commitments;
- providing income – generating scopes so that they can afford to pay for helping assistants necessary for sharing domestic load;
- making arrangement for informal educational or coaching centres where less privileged women may get the chances to be enlightened about their rights and duties;
- adopting participatory methods of learning to make the women aware of different social issues.

Furthermore, it will not be improper to quote the famous saying here that "charity begins at home". Hence, it will be desirable for all concerned male members of the Panchayat (sitting as well as ex-members) to broaden their mind-sets to acknowledge the role of women on the overall development of the society. This proposition may sound utopian but there is no alternative to the change in their own perception and attitudes. This is the only way to reverse the invisible power relations that underlies the social structure even today.

10

Socio-Economic Development of Women

A Case Study of Self-Help Groups in Midnapore District

M.V. RAO

Introduction

Women constitute nearly half of human population in any society. Because of traditional neglect, discrimination and lack of adequate opportunities, most women cannot realise their full potential, achieve individual goals and contribute to the development of society. Rural India exhibits similar scenario of their struggling whole day for daily income and attending to domestic works like taking care of family needs.

An effort has been made by the Midnapore district administration[1] and panchayats to reach thousands of such women to improve their social and economic condition by forming Self-Help Groups of women.

Ground Reality

- In Pingla and Sabang areas in the district of Midnapore, women make mats or *madur* which are famous in India and abroad. They depend on moneylenders and middlemen for small amounts of credit to buy raw materials. The rate of interest can go upto Rs. 10 per Rs. 100 per month. They also depend on these people to sell their products and in the process get paid only a fraction of the market price.
- In Jhargram and parts of Garbeta area, women make rope from *babui* grass. They need only small amounts as capital to buy grass. They are forced to depend on traders for this

small capital. The result is that they get far less price for the rope.

- In many areas, there is a concept of *bhag chas* of goats. A person gives his goats to poor women of the village to rear them. They take back the goats after they become old enough for sale. The woman who rears the goats for months together is paid small amount like Rs. 100 to Rs. 150 per goat. The woman can as well rear her own goats but she cannot think of owning a goat because of poverty and lack of capital.
- Women in forest areas of Jhargram, Salboni and Garbeta, collect *sal* leaves which are used in making plates. They can earn more if they have a pressing machine but they can not afford the same. Women also collect *tendu* leaves in Jhargram areas but the middlemen run away with the cake. These women are not organised and do not have market access and money to withstand the might of middlemen.
- Many women in Daspur manufacture incense sticks or *agarbattis*. The middleman or the trader supplies sticks and the paste. These women are paid paltry amounts as wages on the basis of thousand sticks rolled. The women can as well buy the ingredients and make the product and sell. They can earn many times more in the process. But, lack of small capital forces them to be exploited by others.

Many such examples can be mentioned which clearly show that these women work hard and produce goods whereas the middlemen take away the profit. In many other cases, they pay exorbitant interest and lose their hard earned income.

When These Women Join Hands

What was necessary was to provide motivation so that women organised themselves to form Self-Help Groups for their socio-economic development. Each Group may have about ten members. When a Group is formed, it brings together likeminded women in the neighbourhood. The Group members save small amounts daily from their earnings. This amount can be even Re. 1/- per member per day. The Group members open Savings Bank account with any nearby Bank where they deposit their savings periodically.

It is important that the Group is built and run on mutual trust. For this, it is essential that the Group works in a democratic manner. The members elect a President, a Secretary and a Treasurer from among themselves. The Group members meet at least once a week at a place convenient to them.

Self-Help Groups for Women

Immediate Goals

- Women can save small amounts on a daily basis which grows into a sizeable corpus over a period of time.
- To get small amounts of loans for urgent needs
- Group members can generate income by various economic activities

Long Term Goals

- Increase in self-confidence, seif esteem and leadership qualities of rural women
- Increased focus on total immunization and family welfare
- Primary education for all children, particularly girl child

The Groups provide opportunities for women to re-discover themselves and become aware of their rights and obligations as citizens. In times of massive unemployment and pressure on job opportunities, the groups give them scope to become self-reliant and take up enterprises on their own.

Spreading the Message of Self-Help Groups: A Mission

Once it has been visualised that SHGs are key to empowerment of thousands of women, entire district administration and panchayats have taken it as a mission to reach out to thousands of rural women in village after village to spread the message of SHGs and to facilitate formation of Groups. More emphasis has been given on tribal areas and forest fringe villages.

Orientation programmes for the facilitators and motivators were organised at various levels and repeatedly to make them fully aware of the concept of SHGs. Success stories of these Groups elsewhere have been discussed and problems they may face in the process have been thoroughly analysed. The ways to overcome the likely problems were also worked out in these orientation sessions.

There were people who suggested pilot approach and test the concepts in few places before these are taken up in the whole district. These suggestions were not found acceptable to the district administration as the concept of Self-Help Groups has been very well tested and found to be successful in many parts of the world including the famous experience of Bangladesh Grameen Bank. The idea was to reach more and more needy sections of the society in an aggressive time frame. Goals were set with mutual discussion among the participants of the mission. There were periodic feed back sessions to share the experiences from the field. A system of reward was introduced to motivate the field workers to put extra effort into this exercise.

The campaign was initiated in March 2000 and in a short time, it could become a people's movement and achieve rapid progress in formation of groups. Despite periodic interruptions like elections, floods, the campaign continued to achieve significant success. By December 2001, Midnapore district had more than 25,000 Self-Help Groups, which was a landmark achievement in the State of West Bengal. Table 10.1 shows Block-wise achievement regarding formation of SHGs in the two districts of Paschim Medinipur and Purba Medinipur carved out of Midnapore district on 1.1.2002.

Table 10.1: SHG Status in Various Blocks

Block	*SHG Formed*	*Groups Receiving Revolving Fund*
Midnapore	691	304
Garhbeta-I	822	458
Garhbeta-II	516	285
Garhbeta-III	526	302
Salboni	539	183
Keshpur	898	611
Kharagpur-I	642	340
Kharagpur-II	303	242
Dantan-I	337	112
Dantan-II	322	195
Narayangarh	481	96
Keshiary	458	210
Sabong	561	205
Pingla	333	117
Mohanpur	283	168

(Contd.)

Table 10.1 (*Contd.*)

Block	*SHG Formed*	*Groups Receiving Revolving Fund*
Debra	433	88
Jhargram	827	336
Binpur-I	692	207
Binpur-II	681	263
Gopiballavpur-I	295	65
Gopiballavpur-II	375	137
Jamboni	308	59
Nayagram	522	53
Sankrail	574	197
Tamluk	604	498
Sahid Matangini	293	197
Panskura-I	637	361
Panskura-II	315	152
Moyna	493	411
Nandakumar	637	517
Nandigram-III	630	391
Sutahata	356	179
Haldia	263	237
Nandigram-I	688	465
Nandigram-II	367	293
Mahisadal	387	210
Contai-I	565	408
Contai-II	376	247
Contai-III	578	461
Ramnagar-I	425	205
Ramnagar-II	417	203
Bhagwanpur-I	279	215
Bhagwanpur-II	442	229
Patashpur-I	275	155
Patashpur-II	202	134
Khejuri-I	272	143
Khejuri-II	341	238
Egra-I	617	466
Egra-II	249	116
Ghatal	457	127
Daspur-I	614	280
Daspur-II	617	435
Chandrakona-I	415	227
Chandrakona-II	393	145
Total	**25623**	**13578**

Status: October, 2003.

Source: District Rural Development Cell, Paschim Medinipur and Purba Medinipur districts.

Savings and Loans

Members save small amounts on daily basis. Most of the Groups started with one rupee a day per member. They moved to higher amounts gradually. Many Groups save two rupees per day and there are other Groups whose members contribute three rupees per day. They realise quickly that the money belongs to them and nobody is going to touch it. Higher member contribution helps to build substantial amounts of Group savings quickly. The Group maintains passbook for every individual. Transparency in working helps to build mutual trust among members. Initial hesitation of a member about other members vanishes as time progresses.

To start with, the savings has become a boon for many to meet unforeseen situations at home like ill health and child delivery. The members are able to take small amounts as loans from the Group. They used to go to moneylender for even such small loans. The success of these Groups economically depends on the basic principle of daily savings.

Some people used to argue initially why they can not contribute on a monthly basis. For example, instead of one rupee a day, why not thirty rupees per month. It is difficult for rural poor women to contribute sizeable amount one time. But money collected can be deposited in the bank weekly, fortnightly or monthly as per convenience. A bank may be far off in rural areas and they can't waste time every day to go there only to deposit money.

The economics of Group savings works wonders. If a member saves two rupees a day, the Group consisting of 10 members saves Rs. 3600 in six months. Three members can take loan up to Rs. 1200 against their individual savings of Rs. 360 only. It is important that the Group charges reasonable rate of interest on loans taken by members. The Group decides the rate of interest, which is much below the level charged by the village moneylender. In any case, the interest amount gets added to the Group corpus. It is observed that many Groups charge interest in the range of Rs. 2 to Rs. 3 per Rs. 100 per month.

What do the Group Members Discuss in Meetings?

The Group meets as many times as they need. They have at least one meeting in a week. But it may be possible that the time gap between two meetings may be longer during sowing and harvesting seasons.

The members keep a note of their discussions in a book. There are Groups who charge fine from members who do not attend the meetings.

Study of working of the Groups shows that the members initially discuss savings and money matters only. But as time passes, they start believing in the potential of Group functioning. Any issue that concerns the members is discussed in the meetings. These can be family problems, a woman's husband coming drunk and beating her and health problems of children. Issues that found repeated mention in the meetings were immunisation of mother and child, education of children, particularly that of girls, prevention of early marriage of girls and family welfare concepts. Here, the role of facilitators and motivators becomes important.

Emphasis on Micro Enterprises

The question comes as to selection of activities and enterprises by the Groups. They may work collectively or individually. The members normally take up individual projects and build their assets. The experience of some DWCRA cases shows that insistence on Group assets and sharing of profits among members has not worked always. An individual will be normally more motivated if the asset is personal.

Table 10.2: An Illustrative List of Activities Undertaken by the SHG Members

Sal leaves plates	Sericulture-tasar/mulberry
Agarbatti making	Catering
Silver ornaments	Bidi making
Bakery	Weaving-cotton/silk
Traditional and decorative mats	Vegetables cultivation and Vegetables business
Bamboo products like baskets	Chira/Puffed rice/Pickles/Chanachur/papads
Babui rope and handicrafts	Fisheries and Ornamental Fish culture
Beauty Parlour	Baby food
Banana/Potato chips	Poultry/Goat rearing
Minor irrigation facility	Grocery Shop
Tailoring & Embroidery	Furniture/carpets
Rural Go downs	Generators
Floriculture/Flower bouquets/garlands	Traditional Pottery/Handicrafts
House dairy	Rearing goats and sheep
Horn products	Poultry & duck rearing
Marine Jewellery	Soap making
Pump set repairing	Radio and TV repairing
Toys and dolls	Book binding

The Groups initially take up only those activities with which they are comfortable and whose market is assured. These activities do not need highly technical skills. Income generating activities which get priority are live stock schemes like house dairy, poultry and duck rearing in backyard, rearing of goats and sheep, fisheries, handicrafts which are traditionally made by the members. Thousands of artisans in our villages who make various items need skill improvement, capital and marketing outlets to earn better returns.

Training of SHG Members

It is extremely important that these Groups get necessary support and guidance in the initial phases. Outside intervention should be minimum as the Groups evolve and start realising their potential to work on their own.

The role of motivators and facilitators is vital in the formative stages of groups. Facilitators can be administrative functionaries, panchayat members, health workers, and teachers. These motivators and facilitators are reasonably cautioned to speak in a language and manner comfortable to the audience. Nothing should be said or done that may undermine their confidence.

Some Dos and Don'ts for Facilitators

- *Speak* from their point of view and understand their constraints
- Appreciate their *concerns*
- We need *patience* while talking to SHG members
- Our ideas and vision should be in tune with their *present needs*
- Introduce new concepts in *simple* and *common* words
- Don't frighten them with *English jargon*
- To repeat, use their language

Skill Development of SHGs

Training is imparted to selected member(s) of a Group in accounting basics, maintaining registers and bank relations.

Experienced organisations and institutions like Krishi Vigyan Kendra, Khadi Village Industries Commission, polytechnics and Industrial Training Institutes have been engaged for skill development and upgradation of Group members, Resource persons like master craftsmen. Training modules have been developed through Line Departments like Fisheries, Animal Resources Development, Sericulture, Agriculture and Horticulture. These are of short duration and the training content is prepared in easy–to–understand format and language.

Bank Linkages: Issues

As availability of cheap credit is a constraint, it becomes imperative for SHGs to access bank credit to undertake projects. Absence of credit linkage will hinder the income generation plans of the Group. Efforts have been made to link these Groups with Swarnajayanti Gram Swarozgar Yojana (SGSY). As it may not be possible to link all the Groups with this programme, banks are being pursued to provide them linkage under NABARD Scheme for SHGs. Where the SHG members belong to SC or ST community, SCP/TSP loans are also being sponsored. Convergence of various programmes is being worked out so that maximum number of Groups can get access to capital with less hassle. There are some good examples of bank taking the proactive role of a catalyst in forming and guiding the Groups. But the overall picture needs considerable improvement.

Prospects of fruitful linkage of SHGs with banks face these hurdles:

- Indifference of many banks to Group needs mainly due to previous experience of bad debts or non-performing assets (NPA). Banks have lent money under various Government schemes over the years. Recovery of such loans has not been satisfactory and hence they are wary of further commitments. They tend to view any initiative or a new plan with scepticism.
- Lack of orientation and initiative of bank staff to the concept and needs of SHGs.
- Inadequate staff in rural branches, at present, due to mergers and closures of loss making branches.
- Banks' insistence on collaterals for loans to the Groups.
- Tendency to disburse only Government subsidy portion of SGSY and reluctance to make commitment from banks.
- Discouragement by some banks not to form more Groups as it means more work for them.
- Blocking of Group savings by some banks, as they want to put the Groups under observation to 'test their sincerity and genuineness.'

Bank Linkages: Likely Future Course

A Group feels the need for credit acutely in initial stages as it takes time to build Group fund through savings. As more and more Groups hit the bank blockade, the following scenario is emerging:

- SHGs present a very good business opportunity to the banks. Once the bank misses the opportunity of lending and the Group survives despite this, it will be the bank that will be running behind the Group for business. There are good numbers of Groups who refuse to play to the bank tune.
- NGOs are increasingly playing a lead role in SHG formation and acting as an intermediary between the Bank and the Group.
- Banks are conscious that transaction costs of lending to the Groups are less compared to loans given to individuals.
- SHGs have started to come together to form their own banks. These banks are flexible in their approach and have member friendly norms. This development will gradually reduce the dependence of Groups on formal banking system.

As the concept of SHGs spreads across the country and micro-enterprises of these Groups and members become successful in income generation, banks will soon realise that they need the SHGs more than the SHGs need the banks.

Self-Help Groups and Social Development

Prevention of Early Marriages

Available data of Health department shows that nearly 40 per cent of girls in the district get married before attaining eighteen years of age. Though the average age of girl at the time of marriage has gone up over the years, a sizeable number of them get married and consequently give birth to children before they are biologically ready for this experience.

The women's groups have the advantage of discussing these issues openly and members can learn from the experience of others. It is easy for a Health Worker or an Anganwadi worker (ICDS) to meet these women in their group meetings and explain the messages.

Immunisation and Family Welfare

Immunisation of mother and child has been goal of the State over many years. Problems persist due to lack of sufficient awareness among target audience about these objectives. About 15 per cent of children are still not covered under total immunisation. In addition to the efforts of Health Department, several administrative organisations and private agencies take initiative to reach the needy persons with these facilities.

SHG members discuss immunisation needs of mother and child in their meetings. One thing they point out immediately is 'pulse polio', thanks to the wide campaign over the years. The same cannot be said of other immunisation needs. Malnutrition among pregnant mothers and children is a serious challenge to the community.

Efforts are made to make the members aware of the need for completing the total immunisation cycle for the child. It needs no reiteration that quality of the immunisation is equally important without which the person will live in false sense of security.

The concept of safe period in women has been propagated among the Group members. A simple calendar has been introduced showing the risky days of a menstrual cycle in red colour (Table 10.3). A woman, who wants to postpone childbirth, has to simply mark each day starting from the day her monthly period starts. In the days marked bold the couple can either choose abstinence or the man can use condom. This method can also be very useful for couple planning spacing between two children. The intervention cost is nominal but the effect is far reaching. SHG members are given these calendars so that they can practise themselves; use it for members of family or for awareness generation in the neighbourhood. One can easily imagine the impact on population growth if this simple method is followed.

Table 10.3: A Woman's Month

1	2	3	4	5
6	7	**8**	**9**	**10**
11	**12**	**13**	**14**	**15**
16	**17**	**18**	**19**	20
21	22	23	24	25
26	27	28	29	30

1. Put tick mark ✓ on 1 on the day monthly cycle starts. In the same way, mark each day.
2. If couple meets during the days marked bold, chances of childbirth are high. If a child is not needed now, a man may use condom during this period.
3. The process will be repeated every month as long as couple wants to postpone childbirth.

It is important to ensure that child delivery takes place at the health centre or under the supervision of a trained person. The message is being spread among SHG members and through them, in the community. This will bring down significantly maternal mortality rate (MMR) and infant mortality rate (IMR).

Social Marketing of Health and Family Welfare Materials

The groups have become effective forum for social marketing of essential items like ORS and condoms. In villages, even today the availability and reach of contraceptives like condoms is limited. About 34 per cent of couples have more than three children in the district. Lack of awareness, inhibitions and inaccessibility to family welfare methods contribute to the low performance of family welfare programmes. Enlightened and confident women members of the Groups are able to solve this age-old riddle. Each member is an ambassador of these concepts in her neighbourhood. IUD camps have been organised successfully in many areas of the district during the period from January to March 2003. The campaign was successful even in areas inhabited by people belonging to tribal and minority communities. 'Child by choice and not by chance' is the motto for this successful initiative mainly driven by the participation of SHGs. The Group members have realised to a great extent the need to keep adequate spacing between children.

The utility of ORS and its capacity to save lives is well known. To translate this into a working proposition, SHGs have started storing adequate quantities of ORS packets for use among their families and in the village when the need arises. They have become more conscious now of the susceptibility of children during diarrhoea attack. They ensure that the affected child is given required quantity of water and salts. As a result, many lives are saved and many more are saved from morbidity.

Sanitation

It is common knowledge that many people have to daily face the ignominy of going out to defecate. The problem becomes more acute in rainy seasons. Women have to undergo this shameful experience daily. During interaction with members, facilitators verify the availability of sanitation facility at home. It often happens that some of the members do not have a sanitary latrine. The need to have a latrine and the negative impact of not having one are discussed. It becomes easy to co-relate incidence of diarrhoea and other water borne diseases with lack of sanitation.

Experience shows that it is easy to motivate the Group members and through them, their families to the necessity of having a sanitation facility. Midnapore district has been a pioneer in the country in Rural Sanitation Programme and Nandigram-II block of Midnapore district achieved the distinction of becoming the first block in India to reach total sanitation level. More than 70 per cent of the households have sanitation facilities installed in their homes in the two districts of Paschim and Purba Medinipur districts. There are some areas in the district in the Western part, where the pace of coverage of sanitation has been historically low. In forest areas and areas inhabited by sizeable tribal population, the members of the SHGs are being easily motivated and they are getting the facility installed in their homes. The district is fast moving towards achieving total sanitation coverage in all households-rich, poor, near and far.

What are the Gains So Far?

We come across the following experience when we meet the SHG members:

- Increase in their self-confidence levels and social status.
- The Groups provided the women an opportunity to develop their leadership qualities and managerial skills. It is heartening to observe that many SHG members became Gram Panchayat and Panchayat Samiti members. Some even became Pradhans and Upa-pradhans of Gram Panchayats.[2]
- Women show better understanding of issues like immunisation of child, education of children, women specific problems and their answers.
- Income level of the members has increased because of economic activities taken up from their own savings. Many Groups could obtain credit linkage which helped them in asset creation and significant improvement in standard of living of the members and their families.

- Women have become a part of decision-making process in family and community. The Groups have provided economic security to women and through them, to the whole family.
- The Group serves as a platform for discussing many issues and problems they face day-to-day. These can be problems like children health, impending child–birth in a family, drinking water problem in the area, literacy, sanitation and pulse polio programme. The issues can be endless.
- The Group helps to build cohesiveness among members and in turn in the community. It gives them the confidence and ability to fight collectively on common issues.
- The campaign to form SHGs has become a mass movement in social mobilization. It made significant impact on the approach of community towards women development and their needs. Panchayats and officials came out with a lot of innovative ideas to strengthen and enrich this programme.

The Areas Where Improvement is Needed

Though phenomenal growth has been achieved in formation of Groups, it needs to be followed up with effective monitoring system. This will help them to stabilize and settle down to the task of socio-economic development of the members.

Some of the areas where intervention is needed are:

- Some Groups are slow in internal lending of savings among members. Internal lending helps members to meet immediate and unforeseen needs. It also grows banking habit among them.
- Some of the Group members can afford to save more than Re. 1/- per day but they are stuck in the grove and continue to save at that level. Motivation will help them to understand that more internal savings will help them to take up bigger projects. They will be more self-reliant also.
- Some Groups meet once a month and some, twice a month.

More meetings mean more cohesiveness and collective thinking which will help them to solve a lot of social issues.

- Some of the Groups are irregular in savings. They need proper backing and support in the initial period to tide over doubts and teething problems.
- Some of the Groups have divided the revolving fund received under SGSY equally among ten members. Though they have taken up income generating activities, they need more orientation in future to come out of the equal division mind–set. The Group should give loans to members as per their needs only.
- Some of the Group members need more orientation on health and family welfare matters. It is natural that it takes time to orient them on these issues.

Looking Ahead

Members of Self-Help Groups have proved that they have the potential to transform their lives. Observing some of the two and half lakh confident members of over 25,000 SHGs, one gets the feeling that they are capable of bringing improvement in quality of life for their families and neighbourhood. They seem to tell that there is no problem that has no solution. Study also shows that they are helping in easily tackling many social problems. The concerted campaign in the district has attempted to awaken their collective consciousness. It has also helped to harness their unexplored potential. The Groups offer the institutional framework necessary for convergence of various activities at implementation level. Most of the Government Departments involved in development works and their huge number of field officials can find beneficiaries for their schemes and programmes from the Group members without searching for them. Functionaries of Health, Agriculture, Animal Resources Departments, ICDS workers, and Literacy volunteers can find perfect setting to undertake orientation and implementation of many programmes and schemes.

The concept and utility of self-help and self-reliance effectively bring people out of the dependent mind–set. The Group members are real entrepreneurs in making. Their creativity gets enough opportunity and scope to prosper in Self-Help Groups.

CASE STUDY OF SOME SELF-HELP GROUPS IN GARBETA-I BLOCK OF PASCHIM MEDINIPUR DISTRICT

1. Name & address of SHG	Kantagarh Khanpara Swasahayak Dal, Village Kantagarh, P.O. Tilari, G.P. Kadra Uttarbil, District Paschim Medinipur
Date of formation	04.07.2000
Savings generated as on 15.12.2003	Rs. 35,312/-
Bank credit	Rs. 55,000/- with United Bank of India, Sandhipur branch
Activities taken up	Vegetable cultivation, rice processing, house dairy, goat rearing, Babui rope making, handicrafts from bamboo, kantha stitching
Group members and income generated	1. Aglima Bibi 2. Sapura Bibi 3. Fatema Bibi 4. Hameda Bibi 5. Manowara Bibi 6. Taslu Bibi 7. Sukura Bibi 8. Mamata Bibi 9. Sadeka Bibi 10. Sova Bibi Approximate additional income of Rs. 900/- per member per month
2. Name & address of SHG	Nischintapur Sardar para Swasahayak Dal, Village Nischintapur, P.O. Tilari, G.P. Kadra-Uttarbil, District Paschim Medinipur
Date of formation	28.09.2000
Savings generated as on 15.12.2003	Rs. 28,695/-
Bank credit	Rs. 55,000/- with United Bank of India, Sandhipur branch
Activities taken up	Rice processing, babui rope making, goat rearing, mat making and poultry
Group members and income generated	1. Anhara Mallik 2. Asma Gayen 3. Rasida Gayen 4. Kalo Khan 5. Mafejala Khan 6. Sanuhara Mandal 7. Saira Bibi 8. Najira Sardar 9. Adina Hur Sardar 10. Majnu Gayen Approximate additional income of Rs. 800/- per member per month
3. Name & address of SHG	Nischintapur Gayen para Swasahayak Dal, Villiage Nischintapur, P.O. Tilari, G.P. Kadra Uttarbil, District Paschim Medinipur.
Date of formation	07.10.2000
Savings generated as on 15.12.2003	Rs. 31, 375/-

(Contd.)

(*Contd.*)

Bank credit	Rs. 55,000/- with United Bank of India, Sandhipur branch
Activities taken up	Goat rearing, babui rope making, woolen garments making, rice processing, vegetable cultivation and orchard
Group members and income generated	1. Latifa khan 2. Angura Gayen 3. Asma Gayen 4. Amina Khan 5. Amina Gayen 6. Manowara Mallik 7. Najira Mallik 8. Hurmadina Khan 9. Rousana Khan 10. Purnima Mondal Approximate additional income of Rs. 850/- per member per month
4. Name & address of SHG	Dhadika Gopalan Unnayan Prakalpa, Village & P.O. Dhadika, GP Dhadika, District Paschim Medinipur
Date of formation	22.11.2000
Savings generated as on 15.12.2003	Rs. 26,800/-
Amount of revolving fund received	Rs. 25,000/- United Bank of India, Dhadhika
Activities taken	Vegetable cultivation, poultry, house dairy, rearing of goats and incence sticks making
Group members and income generated	1. Kalpana Mandal 2. Minu Pandit 3. Tulsi Pandit 4. Parul Pandit 5. Karuna Pandit 6. Puspa Pandit 7. Mithu Pandit 8. Latika Pandit 9. Mayna Pandit 10. Uma Pandit Approximate additional income of Rs. 650/- per member per month
5. Name and address of SHG	Mulunia Adibasi Unnayan Prakalpa, Vill. Dhabani, P.O. Dhadika, GP Dhadika, District Paschim Medinipur
Date of formation	13.3.2000
Savings generated as on 15.12.2003	Rs. 30,112/-
Bank credit	Rs. 145,000/- with United Bank of India, Dhadhika
Activities taken up	Rice processing, Vegetable cultivation, rearing of goats, poultry, vegetables business, piggery, transport
Group members and income generated	1. Gobinda Murmu 2. Jadunath Mandi 3. Kebal Saren 4. Satyanarayan Saren 5. Sunil Murmu 6. Kali Charan Saren 7. Chana Hembram 8. Rabin Saren 9. Sanatan Hansda 10. Jiten Mandi Approximate additional income of Rs. 1100/- per member per month

NOTES AND REFERENCES

1. The author worked as District Magistrate and Collector of Midnapore district and later Paschim Medinipur district during the period the SHG campaign was launched and sustained.
2. There are three tiers of Panchayat organisations in West Bengal. The Zilla Parishad is at the district level, Panchayat Samiti at the block level and Gram Panchayat for a cluster of ten to twenty villages. Zilla Parishad, Panchayat Samiti and Gram Panchayats are headed by Sabhadhipati, Sabhapati and Pradhan respectively. The population of a Gram Panchayat varies between ten to twenty thousand and that of a Panchayat Samiti between one to two lakh depending on the size and location.

11

Decentralized Development and Micro-Credit

The Role and Efficacy of Self-Help Groups

APARAJITA DHARA and NITA MITRA

Introduction

In most of the developing countries the banking functions regarding lending activities are undertaken principally by state-owned formal sector banks and partly by some private sector banks. These banks have to operate according to rules and regulations set by the government (which in turn reflect the policies and priorities), and under such a situation, the concept of a profit-maximizing private sector banking system may not seem very helpful. So the formal sector of the credit market is virtually dominated by these government-owned banks.

Now it has been found that in the Less Developed Countries a relatively small part of the population, especially in the rural areas, is covered by the formal banking system – and then also, those covered often tend to be the wealthier and politically influential people. A major problem facing the formal sector banks – which affects their lending activities and which may be a cause of their poor performance by standard economic criteria – is a high rate of loan-default.

The issue of non-payment of loans has drawn the attention of researchers, and the reasons for this may be analysed in terms of the following:

(*i*) Inability to repay.
(*ii*) Unwillingness to repay.

Inability to Repay

Inability to repay a loan may simply reflect a failure of the project undertaken. Apart from the inherent risks involved in agricultural loans, imperfect information on the part of the lenders has been identified in the literature as a possible cause of non-repayment. The problem has been categorized as one of adverse selection and moral hazard. Due to limited availability of infrastructures and inadequate incentive schemes, lenders may be unable to identify good projects and therefore, may have to lend to 'bad' projects (giving rise to the problem of adverse selection) – and may also find it too costly to monitor the project implementation (which brings in the problem of moral hazard). Moreover, for the state-owned formal sector banks, ex-ante profitability may not be of much importance, where lending as an activity often becomes an end in itself (irrespective of any social or private profitability criteria), and so lenders may not be much concerned about the proper utilization of the funds. All these factors together may lead to the failure of the project and the borrower then is unable to repay the loan.

Unwillingness to Repay

The other problem – that of unwillingness to repay the loan – may arise when even if the borrowers have the resources to repay (i.e. possess the 'ability'), they may nevertheless not have an incentive or any compulsion to actually do so. This may partly be due to the difficulties in getting the bank officials to make the effort to enforce repayment -- bank officials, as agents of the depositors and the government, may not have proper incentives for enforcement. Or, despite the widespread lack of acceptable forms of collateral, even where it may exists, when costs of seizure of such collateral are high, effective mechanisms for realizing dues may not be possible.

Another reason as to why such difficulties persist with formal loan programmes may be the lack of political will. It has often been found that the government is not always willing to enforce repayment, due to political reasons. Many programmes thus resemble grants and they become means of making transfers, furthering political rather than economic goals. This is supported by the fact that a disproportionate number of large and influential farmers are defaulters. And when government is not committed towards loan repayment, the borrowers tend to form the opinion that if they can hold out long enough, the loans may eventually get waived. Thus the political environment fuels borrower's expectations that they will not have to repay – and this undermines their incentives to use the funds wisely, worsening the problem of inability to repay.

One may thus say that the major problems faced by the formal sector credit programmes, as far as non-repayment is concerned, are – the dual agency problem and the possibility of political manipulation of the state-sponsored banks.

Due to such high default rates, and some other inherent weaknesses in the banking systems of the developing world the reach of the formal credit system bypasses a large section of the population in these countries, especially in the rural areas. Therefore, one still finds the widespread existence of various forms of informal credit institutions. Ghate (1992) has presented a survey of different informal credit institutions in the Asian context. (There it has been observed that the share of rural informal credit, although declining, still accounts for about 2/5ths of the total rural credit in India).

Informal finance is highly heterogeneous but in all the cases informality is characterised by personalised loan transactions (entailing face to face dealings with borrowers) and flexibility with respect to loan purpose, interest rate, collateral requirements, maturity periods and so on. In order to understand the working of the informal credit market one has to look at the different sources of informal credit prevalent in these countries – the more important ones being: loans from friends, relatives, community members, rotating savings and credit associations, money-lenders and informal banks, tied credit, and pawning.

Thus, it is found that while the formal sector appears to have a comparative advantage over the informal sector in intermediating funds and reaping the benefits of scale-economies, it seems to fare worse in solving the enforcement and information problems. So, in the recent discussions of financial markets in LDCs, there has been an increased interest in trying to combine the advantages of both the formal and the informal sectors. The concept of credit-co-operatives has again re-emerged and group-lending activities have come into prominence of late.

Credit Co-operatives

Credit co-operatives or credit-unions work like the formal sector banks in some respects, raising deposits from and issuing loans to their members. They are often set up with some form of government involvement, and are actively promoted as an alternative to the village money-leanders (who often charge exorbitantly high rates of interest). In large scale banking, even though an individual may wish to monitor the activities of the borrowers, (since high returns depend on the borrowers' performance) it may not be profitable as it is prohibitively costly to do so. But for credit cooperatives this may be possible, given the lender's local knowledge regarding the borrowers and the relatively small size of the membership. This reduces the agency problems experienced in formal banking. Moreover, to check the moral hazard problem, the non-borrowing members of the co-operative may be made creditors and some of the funds be raised internally from the members themselves. But problems may arise when there is a large number of non-borrowers and if no non-borrower wishes to put in the extra effort of monitoring the proper utilization of the funds, the problem of free-riders may prevail. Also, it should be ensured that the co-operatives remain collusion-proof, so that members (having close information regarding fellow members) do not form groups within the co-operative itself. These may be some of the reasons as to why the performance of co-operatives in reality falls short of expectation.

Group-lending programmes

Besides co-operatives, group-lending programmes have also gained much popularity in recent times as a way of improving the quality of loans from formal sector institutions to poor individuals. Such group-lending activities try to reap the advantages of local knowledge, as well as that of peer-monitoring, where fellow members of the group are given some liabilities, so that they are compelled to monitor the activities of other members of the group. Thus, members of the group themselves provide the information and discharge enforcement functions. Individuals may also repay each other's loans, if the project of any member fails, and this encourages mutual assistance. The Grameen Bank of Bangladesh, formed by Md. Yunus, is often cited as a paragon of such institutional design. But such a scheme also has some drawbacks. It cannot guarantee against the possiblity that 'bad' borrowers do not inflict negative externalities on 'good' ones – a 'good' borrower might be prepared to repay his own loan but might not be sanctioned loans anything, if a partner in his group does not repay. This may negate the positive effects, if the clause is made such that the group as a whole is denied loan in the next period in case of a default. Research studies, as by Huppi and Feder (1990), reveal a mixed performance as far as group-lending programmes are concerned, and we are unable to generalize that it leads to higher rates of loan repayment than conventional lending programmes. There are various aspects regarding these programmes, which require further detailed investigations.

Thus, it may be observed that due to some inherent economic reasons as well as political problems faced by the formal credit institutions, a large section of the rural population in developing countries, still depend on the informal sector. Being not integrated to the formal sector, the scope of exploitation (in the informal sector) may not be reduced via state regulations and policies. So efforts are being made to combine the advantages of both the sectors, and to design alternative institutions of co-operation, group-lending, etc. which may integrate the positive features of both.

This paper makes an attempt to focus on the working of some Self-Help-Groups in order to understand the extent of advantages that may be enjoyed when micro-finance institutes, having both formal and informal credit market characteristics are set up.

Self-Help Groups: Some Concepts

Group lending schemes, that put local knowledge to use, have a high potential of overcoming the problems of 'imperfect information' faced by the formal credit institutions; especially regarding loan repayment. These are the problems of:

(*a*) Involuntary default (stemming from adverse selection as well as moral hazard).
(*b*) Stregetic default (due to moral hazard).

Stiglitz' study (1990) focuses on the informational advantage of group-lending – the fact that the group members may have better information than the formal sector banks about the borrowers' abilities and efforts, increases the likelihood that the borrowers will be able to repay a loan (assuming, of course, that one will not default voluntarily). On the other hand, Besley and Coate's paper (1995) analyses the borrowers' willingness to repay the loan after some set of project returns has been realized, and how group pressure may help to increase that willingness.

A key feature of group-lending schemes is *joint liability.* One way of encouraging members to repay a loan is to require some mandatory deposits that are reimbursed only when all the borrowers repay their loans, but the practice of denying credit to any of the other group members in case of default by any single member, is the most effective and least costly way of enforcing joint liability. All group members are treated as being in default if any one member of the group does not repay his/her loan, and until the debts of the group are discharged, no member can gain access to further credit. Joint liability thus induces interdependence among borrowers, and acts as an effective incentive device for monitoring the performance of co-members.

Group members generally have local information about each other, and this may help in successfully keeping the potentially delinquent borrowers – whose ability and intentions the members are not so sure about – away from the group. Thus, would-be members who are considered 'bad' credit risks would be excluded, leading to better targeting of borrowers, and the probability that a borrower will be 'good' will increase. The problem of *adverse selection* – that is the selection of a 'bad' borrower because of imperfect information – may thus be taken care of to a large extent. Moreover, due to the threat of future denial of loan in case of a default, the members have the incentive to urge each other to take on safer projects, and also monitor that the loans are utilized for the purpose it was taken and not frittered away for wasteful consumption (i.e. a case of *moral hazard*). Thus, *peer monitoring* under group-lending schemes can reduce the probability of involuntary default (or 'inability to repay').

Peer monitoring may also ensure that the loans get repaid and may be an effective institutional arrangement to overcome the problem of strategic default. This is because, under individual lending contracts, all that the borrower has to fear in case of default is the penalty that can be imposed by the banks. However, under group-lending the defaulter may also incur the wrath of other group members. Since the cost of upsetting other members of the community may be quite high, especially for groups that are formed from communities having a high degree of social cohesion, peer pressure may also help in lowering the probability of 'wilful' or strategic default.

Another effect of joint liability on repayment is that successful group members may have an incentive to repay the loans of a co-member whose project has not yielded sufficient returns to make repayment worthwhile. This aspect has been analysed in Besley and Coate's (1995) article. But, here a negative effect may also arise. Group members have little incentive to repay in majority of their peers default. There may be cases when most of the members' projects have failed and the group has a high chance of defaulting as a whole. Under such a situation, some members who would have repaid under individual lending may choose to default voluntarily, since they know that even if they repay, the group will be treated as a defaulter. But, unless the activities of the members are highly correlated and unless most of the members are given loan at the same time, the risk that the whole group will default is quite low. Thus, the positive features of joint liability generally overcome its negative sides, especially if group cohesion is strong and members feel responsible for the effect of their actions on others.

Apart from joint liability and peer pressure, another important characteristic of successful group-lending schemes is the small size of the groups. Although a small size raises the risk from a single member's default, it increases the incentives for peering monitoring. With larger groups there may arise the *free rider* problem—each member would prefer the others to expend the required energy for monitoring and incur the ill-will resulting from reporting the offenders who have misused the loans. Moreover, as the group size increases, the costs incurred by an individual member as a result of default by any other member may become sufficiently small so that the incentive to monitor is minimal. Thus, for the proper functioning of the groups, they should be of some optimal size.

One may thus conclude that, informational and enforcement advantages (through peer monitoring), as discussed in the current literature, have, of late, led to the evolution of self-help small savings groups, in many of the less developed economies. The leading example of small scale lending (or micro-finance) is the Grameen Bank of Bangladesh. In many other LDCs also—like Thailand (the Bank for Agriculture and Agricultural Cooperatives—or BAAC). Indonesia (The Village Banks system of Bank Rakyat), Malaysia, Nepal, Malawi—experimentation has been going on along these lines. They are of different types, but a common element underlying all these systems is their evolution as a response to overcome the deficiencies of the formal credit market by bringing in the advantages of informal finance.

The Indian Experience regarding SHGs

Limited Reach of the Formal Credit Sector in Rural Areas

The nationalization of commercial banks in India has seen the number of rural branches increasing from 1832 in 1969 to 32498 in 1989 – constituting 57 per cent of the branches of the banks. But, despite such a vast expansion of the formal credit system, it has been found that the rural poor – especially the socio-economically backward sections like landless labourers, marginal farmers, small traders and local artisans—still depend largely on informal credit sources like the moneylenders and pawnbrokers, in times of need and has to pay high interest rates. Owing to difficulties faced by the banking sector in dealing effectively and economically with a large number of small borrowers, requiring credit frequently and in small quantities, and also owing to the banks' perceptions of the credit-worthiness of such borrowers, the credit flow to these sections of the population has remained uninstitutionalized to a great extent. Moreover, when the economic operations (of the poor) are at a subsistence level and the line dividing 'consumption requirements' and 'production needs' is rather thin, it is difficult to adopt the project-lending approach of the banks.

Voluntary Associations and the Tradition of Group-savings

Under such circumstances, several Voluntary Associations (VAs)—working for the development of the weaker sections, through informal education projects, health programmes, skill-improving training schemes, etc.—have encouraged the rural poor to form small-savings groups and practise thrift. The VAs have taken initiatives to inculcate a savings habit in the group members, so that in times of emergency they may have some sort of a savings cushion to fall back on, and if necessary, may also take small loans from the group's savings.

Group Approach under the DWCRA Scheme

The group-approach was also adopted for implementing the government-sponsored DWCRA (Development of Women and Children in Rural Areas) scheme—a sub-component of the Integrated Rural Development Programme—to bring women together for pursuing economic activities. There also it was found that the Non-Government Organizations promoting the DWCRA groups have encouraged thrift and small savings among the group-members.

The common characteristics of all these groups, whether formed individually or under any programme, are stated below.

- The groups are homogeneous entities—formed by people belonging to the same locality, and having a more or less common economic and social background.
- Each month the group-members save some small amounts and in the periodic group-meetings (usually held once or twice in a month) deposit their savings with the group-leader.
- Short term and small intra-group loans, for both consumption and production purposes, are made available to the members, depending upon the need and available resources.
- All decisions regarding the group-activities are taken democratically by active participation of the group-members.

Co-operatives versus *SGHs*

The earliest form of such Self-Help-Groups was the Co-operatives Credit System which aimed at bringing together poor people and promote mutual self-help concepts through small savings. But over the years, owing to the presence of vested interests many co-operative institutions became bureaucratised and politicised to safeguard the interests of a few. SGHs have better chances of becoming successful because of their highly cohesive nature. Typical SHGs consisting of poor members having homogeneous socio-economic background are likely to serve the credit needs of the poor better, since they are managed through consensus of all members in the periodic meetings.

Resource Base of the SHGs

It has been found that while small size gives a participatory nature to the group, it also results in the disadvantage of a very small resource base; as the members are all from poor families, the amounts they can possibly save is very little. Often the concerned NGO supplements the SHG's resources by providing loans from their own funds. However,

VAs generally run on donor-assistance, which is not much and also quite uncertain. Their capacity to support SHGs financially is, therefore, limited.

Advantage of a Linkage with the Formal Sector

It can thus be seen that, from the SHG's point of view, a link with the formal banking sector may enable the group to have access to a larger quantum of resources and enhance the operational scale of the group-members.

From the bank's perspective also, if proper links can be established with such SHGs, bank financing could lead to lower transaction costs and systematic credit supervision.

NABARD Project for Developing a Bank—SHG Linkage

To develop a supplementary credit delivery system for reaching the poor in a cost effective and sustainable way, the National Bank for Agriculture and Rural Development (NABARD) has introduced a pilot-project in 1992, for linking up SHGs with banks.

The broad principles followed under the linkage scheme are:

- Savings should be done first and there will be no credit without savings.
- Keeping savings as partial collateral, banks will give loans to SHGs for on-lending to group members.
- Joint liability should act as a substitute for physical collateral.
- Interest rates and other terms and conditions for loans will be decided by members themselves.
- Ratio between credit and savings is contingent upon the credit worthiness and could increase over a period due to good repayment record.
- To begin with, small loans will be given.

Various studies were conduced by NABARD to review the linkage programme and the findings confirmed that

(*i*) membership for the SHGs mainly came from poorer sections of the population and there is large scale participation of women;

(*ii*) even the very poor are able to save and their savings increase with increase in income;
(*iii*) demand for credit is frequent and for small amounts and not necessarily for the purchase of income-generating assets;
(*iv*) SHG intermediation leads to lower transaction costs for both the banks and the borrowers and
(*v*) due to group dynamics, recovery of loans is nearly 100 per cent.

The linkage programme proved beneficial to both the banks and the groups, in terms of reduction in the transaction costs.

It is often the case that the banks understand the importance of mobilising tiny savings but find it prohibitively costly to service the high volume of small accounts. Under such circumstances, SHGs may fulfil the role of the own-present savings collector, the members pool their savings and the group deposits the pooled sum with the bank. Similarly, some of the costs involved in dealing with a large number of borrowers' accounts also gets externalized when lending is done through groups. The banks, therefore, can successfully cater to the requirements of the poor while cutting their transaction costs on three accounts:

- They have to handle a larger group-savings rather than larger number of small individual amounts;
- They save on the appraisal costs of non-viable cases, which are not forwarded by the NGOs. The appraisal by the VAs provides the bank with adequate material for their own appraisal.
- The VAs help in monitoring and recovery of the loans. VAs may also help in providing the borrowers with necessary training in asset-maintenance.

Such encouraging findings prompted NABARD to extend the linkage project beyond the pilot phase and promote SHG-lending a regular banking activity.

(But, here it may be observed that since the entire credit cycle of need assessment, appraisal, disbursement, monitoring, supervision and recovery is being undertaken by the concerned NGOs, some portion of the transaction costs actually gets transferred from the banks to the

NGOs. As long as the SHGs remain dependent on their parent NGO, this may continue to be the case).

Beside the NABARD linkage scheme, SHG-lending is also encouraged under the government's Rashtriya Mahila Kosh scheme (National Credit Fund for Women under the Department of Women and Children, Ministry of Human Resource Development). One of the objectives of this fund is to provide low-interest loans to women from below-poverty-line families, through groups.

Thus, we find that SHG-lending has a high potential of becoming a popular and successful means of extending the reach of the formal credit sector in the rural areas, especially for the poorer sections of the population.

A Case study

In this paper, we try to evaluate how SHGs are faring considering the district of Howrah in West Bengal state as a case study. For locational advantages, and also because of the presence of some well-established Voluntary Agencies engaged in social work and public service, this district has been taken up for the study on the workings of Self-Help-Groups. The hypotheses that our study has tried to examine are:

1. SHGs are an effective mode of savings mobilisation and credit delivery to the rural poor.
2. SHGs play an important role in improving the coverage of rural poor by the formal banking sector.

For this purpose, a case study of some sample SHGs in Howrah district has been made. The research methodology involved mainly primary data collection from the Non-Governmental Organisations and Voluntary Associations promoting such SHGs, through discussions. The information thus collected was corroborated by attending a group meeting of a particular SHG, where the views of the group-members were sought to facilitate comparison with the information collected from the voluntary sector. Open-ended questions in the form of an interview with the group-leader and the members were administered. It was our purpose to avoid the use of structured questionnaire, as it was felt that a lot of vital information may be left out in the process. The interview-responses were arranged in an informal setting and were recorded verbatim. The information set was further enlarged by the collection of additional data from the officials of the district Lead Bank, the NABARD, and the commercial banks concerned. Apart from primary data, secondary data were also obtained from official reports and other publications.

However, the information gathered could not be subjected to any strict statistical analysis, largely owing to the following two factors:

(*i*) Small size of the sample set, and
(*ii*) Subjective and qualitative nature of the information set.

Using the methodology the study has been kept confined to some SHGs in the blocks of Panchla, sankrail, Uluberia, Amta and Jagatballavpur. The selection of these blocks was made in consultation with the bank officials conversant with the financial implications of SHG on priority sector lending.

The findings of our study are presented below and certain policy issues are discussed.

Findings

(a) *Group Formation*

- The SHGs surveyed under this study were found to have been formed as part of women's development programme and, therefore, all were exclusively women's groups.
- The target group for these development projects was the poor and the destitute women and preference was given to women belonging to the socio-economically backward sections of the population. Thus, most of the group-members were found to belong to below-poverty-line families.
- The groups are all locality-based and the members have social connectivity with each other.
- For selecting the appropriate target group, the NGOs interacted with the local level Panchayat functionaries as also village level organizations and youth clubs.

(b) *Group Characteristics*

- Membership in the groups was found to be between 12 to 25, the average group size being 20. The NGOs were of the opinion that the groups could not function properly if membership fell below 10.

- The women belonged mainly to the age group of 18 to 50 years (when the productivity of a person is high) and most of them were found to be around 30-35 years of age. This appears to conform to the target age-group of family-planning, literacy and other social development programmes that are vital for the uplift of women in particular and society at large.
- At the time of joining the groups, around 50 per cent of the members were illiterate—but they were given some education by the NGOs and many of them were made literate over the years. At present, around 15 per cent of them is still illiterate.
- In many groups, the members received some training in areas like *zari*-work, foam-leather bag making, jam preparation etc., provided by the NGOs. This helped in raising their level of skill.
- It was observed that the members do not carry out a common 'group-activity' as such, though a large section of the women are involved in similar economic activity, i.e., zari-work. This was because the zari-industry is a predominant home-based industry in the block of Panchla, where most of the groups were located.

(c) *Group Activities Regarding Thrift and Credit Programmes*

- The groups practise small-savings and credit programmes.
- Group meetings are held fortnightly or monthly in a member's house (generally in the afternoon), and the savings of the members are collected.
- The group members select one leader and one cashier from themselves, who they feel to be trustworthy and capable. To enhance the participatory nature of the group, and to increase the managerial capacity of all the members the leadership generally changed from year to year. This change has prevented the concentration of power in the hands of a few, and malpractices by them.
- The members were found to save, on an average, 20 to 50 rupees per month.
- Almost all the groups have opened savings account with the

local banks and these accounts are operated by the group leader and the cashier and, in some cases, the parent NGO is also the signatory.

- Members are given short term and medium term loans from the group's savings, for both consumption and production purposes. Production loans are found to have received preference over consumption loans and the amount of consumption loan is generally less than the former.
- There is minimal documentation between the borrowing member and the group, to simplify the lending procedures.
- Out of the groups surveyed, only 20 SHGs (under one NGO) were found to have taken loans from the formal sector banks. The other groups are carrying out only the savings programme with banks and have not gone for any credit-facilities from the commercial banks or RRBs. Most of them are, however, carrying out the transactions with banks on a regular basis and are eligible for loans in future.
- In some cases the group members also availed themselves of loans from the NGO's own funds.
- The interest rate charged on the intra-group loans varied from group to group, as decided by the members themselves. The usual practice has been to charge around 15-20 per cent, on loan amount which is significantly lower than the informal sector interest rate.
- All decisions regarding sanction of loan, loan amount, period of repayment and the interest rate to be charged are made democratically in the group-meetings.

(d) *Repayment Performance*

- The repayment rate for the intra-group loans was observed to be around 95 per cent, and cases of default (especially strategic default) was very low.
- The survey found the major incentives working towards this high rate of repayment to be:
  - — Peer pressure, which resulted from a joint-liability condition as well as the presence of a savings component in the loans;

— The sense of belonging to a group and the realization of its benefits;
— A surety that if the current loan is repaid in time, the borrower may avail of loans again in future. This guarantee of getting loans again, and in times of need, was found to work as a strong incentive for repaying the loans according to schedule.

(e) *Problems Faced*

- It was observed that the members are, at present, emotionally highly dependent on the promoting NGOs. They do not feel confident enough to carry out the thrift-credit activities on their own, without the umbrella-protection of the Voluntary Associations. Thus, the members cannot be said to be hard decision makers, and their process of empowerment is currently at an elementary level only.
- This dependence was more prominent in case of maintenance of the group-accounts. Although the procedures are quite simplified, and the leader and the cashier are given some training, they are finding it difficult to maintain the accounts-books and seek out the NGO's help. This is one area where the NGO's guidance will be required for some time to come.

(f) *Benefits to the Group Members*

The benefits that accrued to the group members for being a part of the SHG are manifold, relating to various economic as well as social aspects. These are:

- Group-savings are found to have imposed discipline among the group members in developing a savings habit and enhance self confidence of the members.
- In times of need, the members can now fall back upon their savings and not to go to the pawn-broker or money-lender (and get exploited by their high rate of interest).
- The loans used for production purposes helped in supplementing their household income to a good extent. Many women were earning around Rs. 1000 to 1500 per month,

from their newly developed activities. This has also led to an improvement in the status of the women within the household (being an earning member now)—and made them more confident about their abilities. Thus, SHGs have a great potential as a vehicle for women's empowerment.

- The flexibility of the group-savings programme has made it more approachable than banks where timings are fixed.
- The involved NGOs, beside facilitating the formation and organization of the SHGs, also discuss various social and economic problems. As awareness creation and non-formal education are integral to the activities of any voluntary association, SGHs (under NGO) not only provide the members with an opportunity to carry out economic activities, but also discuss and analyse their overall situation to arrive at the root cause of their problems and find solutions by themselves.
- There has thus been an all-round improvement in the health and hygiene and general education level of the women. The members are spending a greater part of their enhanced income on children's education and all of their children are attending schools. Thus in the long run, SHGs may indirectly help in improving the area's Human Development Index to a great extent.

(g) *Views of Bank officials, NGO officials and Group members regarding SHGs and the Linkage Programme of NABARD*

Although most of the SHGs covered under this study have not taken any loan from the formal sector banks, the interviews with bank officials, NGO workers and the group members revealed the following:

- The banks prefer the linkage with SHGs through NGOs. They find it more secured to give loans to a registered NGO for on-lending to groups, and are not that much confident about the direct SHG-Bank linkage as yet. (Moreover, since there has been only a few instances of credit-linkage with SHGs, the bank officials are viewing such linkage only as a special lending facility and not as a regular banking activity).

- The NGOs, on the other hand, are in favour of the direct Bank-SHG linkage. NGOs do not want to get involved with loan-distribution as this will increase their operating costs and also because of shortage of manpower. The field mobilizers and coordinators were said to be overworked with other developmental programmes. Moreover, some NGOs were of the opinion that this may have a negative impact on their development-linked image if they get involved with loan-related matters. The NGOs want to provide guidance to the SHGS but not any direct involvement in the linkage programme.
- The Self-Help-Groups, however, would like the linkage to be through their parent NGO, as the members are not yet confident enough to carry out the transactions on their own. It was found that the concept of SHG-Bank linkage is still quite unknown to the people, and they are not much aware of the procedures.

From the above field-level study it can be said that though the SHG-Bank credit linkage is not much prevalent in the study area, the idea is slowly catching up with the people in general and women in particular. SHGs may thus play an effective role in mobilizing rural savings, and have huge potential in increasing the coverage of the so-far unreachable rural poor by the formal credit institutions. By forming such groups, the poor can not only gain access to the formal credit sector, but also improve their economic and social conditions.

Policy Issues

To make the Self-Help-Group mechanism viable and effective certain policy issues may be discussed.

- *Group Formation.* The credit groups should be allowed to come up on their own or through the intervention of voluntary agencies, and not by the credit institutions for giving loans. If the group takes shape on their own and at the initiative of the members, interaction and group dynamics may foster group-consciousness and ensure solidarity.

- *Savings Component:* For self-reliance and reduced dependence on external funds from financial institutions, a strong savings programme may be helpful. Savings may help the members in not taking loans for every small purpose. Moreover, the existence of a common pool of fund may also enable the members to acquire expertise in fund management. This experience may encourage the group members to take up larger projects in future. Moreover, groups that mobilize savings deposit, may increase the borrower's incentive to repay.
- *Need for External Source of Finance:* Thrift based SHGs have shown that given the flexible conditions, the very poor can also save something. But unless a lending programme is also set up, the income generating capacity of the members may not increase much. Loans at consessional rate in the initial years from the formal sector may be useful for SHGs to augment their resources, and once the process gets going, loans may be given at the competitive rate.
- *Frequent Availability of Credit:* Contrary to the target-based poverty alleviation programmes where one-time credit assistance is provided, it may be envisaged that an SHG should cater to the frequent credit needs of its members for sustainable development, and the banks are expected to provide more credit as the performance of the group improves.
- *Flexibility:* Too much flexibility in the operation of SHGs may hamper their growth and prevent the replication of the model elsewhere. In countries like Bangladesh, Indonesia and Malaysia, where SHGs have been found to be very much successful, a rigid and inflexible system is followed and no freedom is given to change any parameter. Therefore, it may be suggested that on certain matters like minimum documentation and accounts maintenance at the NGO/SHG level the system could be a rigid one, allowing flexibility of operation to individual SHG members. This may help in getting some standardization of procedures, and ensure that the NGOs do not make mistake or incur huge start-up costs.
- *Presence of NGOs:* Lending through a registered NGO may give a formal characteristic to SHG-lending and is found to

be preferred by the banking sector. But the NGOs and VAs (promoting SHGs in the rural areas) were found to depend to a large extent on the honesty and trustworthiness of both their staff and the borrowers. SHG operation also puts a tremendous work-load on the field level officials. At present there is considerable interest in the developmental programmes, but in the absence of a strong incentive structure, futurte productivity of the staff members may fall.

Moreover, since the groups are found to be very much dependent on the NGOs, there is a need for more NGOs, sincere and dedicated to the cause of the poor people—especially in the initial years.

Future Study

It may be observed that two possible situations may arise:

(*i*) SHG members pursue their own individual activities, without any common element among them.
(*ii*) All members may pursue a single activity in a co-operative manner.

The question that now becomes interesting from the economists' point of view is whether the survival rates of these two types of groups are same or there are any differences.

So, the next area of enquiry would be to see whether the quit rate is higher for those SHGs where activity mix is heterogeneous as compared to those where only one productive activity is jointly pursued by all members.

The answer to this question may be very important particularly in the context of the past experience with the cooperative sector and also the future stability of SHGs.

Under the newly introduced (from April 1999) government programme—Swarna Jayanti Gram Swarojgar Yojana—which merges some existing poverty alleviation programmes like the IRDP, DWCRA, TRYSEM etc., more stress is being given to group financing. The objective of the government is to take up a group approach under SJGSY, and under this programme many new SHGs are going to be formed throughout the country. Steps will be taken to nurture these groups and enable them to function effectively. Group activity will be given preference and a major share of the SJGSY assistance will be in activity clusters. However, around 25 per cent of the assistance may be made available for other activities and to provide flexibility. Credit will not be a one time affair any more, and the beneficiaries will be encouraged to increase their credit intake over the years.

SHGs may also be promoted by institutions other than SJGSY. Since micro-credit institutions and SHGs are found to be important vehicles for delivery of credit to self-employed people in rural and semi-urban areas, RBI has attached particular importance to the work of NABARD and public sector banks in this area. It has been mentioned in the Monetary and Credit Policy for the year 1999-2000, that a special cell manned by a senior officer from the commercial banking sector with practical experience in micro-credit is being set up in the RBI in order to smoothen the flow of credit to SHGs.

If such points are taken care of, SHG-lending may become a successful credit delivery scheme, especially for the economically and socially backward sections of the population.

Thus, it may be concluded that in the present context of economic reforms this public-private partnership (and the innovative institutional arrangement where both the local population and the voluntary organizations may be involved in the decision making process) may help the process of empowerment of the poorer sections of the society.

NOTES AND REFERENCES

Besley, T. and Coate, S. (1995): Group Lending, Repayment Incentives and Collateral, *Journal of Development Economics,* Vol. 46, 1995.

Ghate, P. (1992): *Informal Finance: Some Findings from Asia,* Published for the Asian Development Bank, OUP, 1992.

Huppi, M. and Feder, G. (1990): The Role of Groups and Credit Cooperatives in Rural Lending, *World Bank Research Observer,* Vol. 5 No. 2, 1990.

Stiglitz, J.E. (1990): Peer Monitoring and Credit Market, *The World Bank Economic Review* Vol. 4, No. 3, 1990.

12

Role of Voluntary Sector in Rural Development

SANKAR MAJUMDAR

Introduction

Human societies have many facets: social, economic, cultural, religious, political etc. Members of the society, either as an individual or as a loosely formed group or as an organized group, work for advancement of themselves and for that purpose they contribute time, energy and resources. With the growth of human civilization and formation of modern nation-state, responsibilities of several such works gradually came under the purview of the state. But even today the state has not taken up (or the society has not entrusted the state to undertake) the responsibilities of all aspects of human societies. Historically, over time the state has been accepting more and more responsibilities. Very recently with the advent of liberalization, the state is giving up more and more responsibilities.

The scope and nature of responsibilities and types of works to be undertaken by the state is determined mainly on the basis of the perceptions of the representatives of the ruling section. In forming their conceptions they are assisted by the technical experts. In the countries of Asia, Africa and Latin America (the so-called 'under-developed' third World Countries) they are assisted by the donor countries and donor agencies. All of them are trained in the paradigm of 'development' – a pivotal concept from 'West' as a uniform prescription to solve the problems of the third world countries (as conceived by the people of the West). A large number and variety of development policies, programmes as well as agencies and administration came forth to address all parts of third world life.

But poverty increased, unemployment proved resistant to growth and food situation could not be solved. Attention got shifted to removal of poverty and injustice, provision of basic needs, empowerment of women, environment etc. New indicators of 'development' were prescribed. These are, for example, malnutrition, illiteracy, infant mortality, life expectancy etc. Here again, the standards were of uniform type and those of western countries.

It was difficult, if not impossible, for this centralized group and their outside think-tanks to conceive the true nature of the needs of the thousands of tiny human settlements in these third world countries characterized by immeasurable diversity.

As a result, a large part of the societal needs as perceived by the people themselves either as a individual or as a group remained to be cared for by the people themselves. They themselves in all these societies do undertake several such works towards social, cultural, economic, political and religious purposes which they consider as essential for living in their society as per their own perceptions of 'Civil Society' and life. For performing these works the people themselves contribute time, energy and resources (human, physical and financial). These contributions are, in some cases, voluntary and in some other cases, these are social impositions. Peoples' association, mass organizations, clubs, villagers' organization, panchayat, political associations and political parties etc. all are based on voluntary involvement of the people. The people as members of the society contribute voluntarily in many such activities.

Hypotheses

Following Hypotheses can be made regarding the role of this 'voluntary sector':

(*i*) All types of rural people do voluntarily contribute towards their own societal works.

(*ii*) People's voluntary contribution is more where the benefits from such activities directly accrue to them and where they themselves are the decision making authority in all the stages of work (planning, implementation and evaluation).

(*iii*) People do contribute to all aspects of rural life.

(*iv*) People voluntarily accept those responsibilities which government does not normally undertake.

Empirical Findings

This section presents two case studies on voluntary contribution of individual rural families. This is based on primary data collected through questionnaire and household survey method from two villages, namely (*i*)Kalyanpur village under Kalyanpur Gram Panchayat or Nandakumar block in Midnapore District and (*ii*) Gurepara village under Sarenga Gram Panchayat of Sarenga block in Bankura District in West Bengal. Selection of the villages were purposive. Two sample villages were the homeland of two M.Sc. final year students of the Department of Economics with Rural Development, Vidyasagar University. Students being the local residents, it was expected that they knew the details of the social, cultural, political, religious, economic, infrastructural activities of their own villages and also about the involvement of their fellow neighbours in these activities. They are supposed to know in a better way the social, economic and other conditions of their fellow neighbours. These aspects are definitely important for getting reliable and authentic information for understanding the qualitative aspects of the problems under present study.

Type and Nature of the Sample Families

All the families were grouped into six categories based on the size of land owned by them. A sample of 60 families from each of the two sample villages was selected using proportional sampling technique. Majority of the villagers in both the villages are marginal and small farmers (Table 12.1).

Table 12.1: Distribution of Sample Families in Two Sample Villages

Categories of Families (ownership Holding)	*Kalyanpur*			*Gurepara*		
	No. of families	*%age*	*Area owned (in acres)*	*No. of families*	*%age*	*Area owned (in acres)*
0.00 acres	10	16.67	0.00	7	11.67	0.00
0.01-2.50 acres	17	28.33	17.55	43	71.67	48.31
2.51-5.00 acres	18	30.00	63.57	6	10.00	23.24
5.01-7.50 acres	9	15.00	55.02	2	3.33	14.16
7.51-10.00 acres	4	6.67	31.50	1	1.67	10.00
10.01+acres	2	3.33	26.54	1	1.67	12.80

Source: Primary Survey, 2000.

Table 12.2(a) and Table 12.2(b) reveal that the majority of the families were dependent on agriculture. However, a portion of the village families was engaged in multiple economic activities.

Table 12.3(a) and Table 12.3(b) show the income pattern of these families. Per family annual income varied from Rs. 33,500 to Rs. 2,41,000 in Kalyanpur. But average annual income of the sample families was Rs. 79,087 in Kalyanpur. The average per family income was much less in Gurepara village (Rs. 42,662). The differences in per family income was also more in Gurepara. It varied from Rs. 14,924 to Rs. 2,45,000.

Voluntary Contribution of the Individual Families

Information on the contribution of these sample families towards different aspects of rural society and life was collected. The aspects considered in this study pertain to different aspects of village people's life such as religious, political, cultural, infrastructural, panchayat etc. Information was collected mainly on the monetary contribution of these families. It may be noted that the rural people do contribute in terms of labour time and physical materials for their own area development. Sometimes they do it on their own and sometimes these contributions form a supplementary part of a programme or scheme initiated by the government and panchayat. Physical contributions made by the villagers have been evaluated in money terms. However, this study has not considered the time and energy contributed voluntarily by the individual rural people for various types of developmental works in their areas. For example, they do spend huge time and energy on making the literacy programme, sanitation programme etc. a success. Another area where they do contribute is raising political awareness of the people through their voluntary involvement in political parties and also through their involvement in the panchayats (local self government).

Table 12.4(a) and Table 12.4(b) present the information on the voluntary contribution of the various categories of sample families in the sample villages. These tables also contain information of the gram panchayat (local self government) tax paid by these families. On an average, individual family's annual voluntary contribution towards various aspects of rural society was Rs. 1786.00 in Kalyanpur and Rs. 384.99 in Gurepara.

Table 12.2(*a*): Distribution of Sample Families According to the Occupation of the Head of the Family in Kalyanpur in Midnapore District, 1999-2000

Land Holding Groups	*Total No. of Families*	*Occupation of the Head of the Family*							
		Agriculture	*Business*	*Service*	*Daily Labour*	*Social Work*	*Service & Social Work*	*Agrl. & Business*	*Others*
0.00	10	—	2	—	5	—	—	—	3
0.00–2.50	17	2	4	1	7	—	—	—	3
2.51–5.00	18	3	2	5	1	1	1	1	4
5.01–7.50	9	3	5	—	—	—	—	1	—
7.14–10.00	4	1	—	3	—	—	—	—	—
10.00–1.00	2	1	1	—	—	—	—	—	—
Total	60	10 (16.67)	14 (15.00)	9 (23.33)	13 (15.00)	1 (1.67)	1 (1.67)	2 (3.33)	10 (16.67)

Source: Primary Survey, 2000

Table 12.2(*b*): Distribution of Sample Families According to the Occupation of the Head of the Family

Landholding Groups	*No. of H.H.*	*Agl.*	*Agl. + Business*	*Agl + Service*	*Agl + Business + Service*	*Agl + Daily Labourer*	*Daily Labour in Other Areas*
0.00 acre	7	—	—	—	—	1	6
0.01-1.00	27	5	1	—	—	15	6
1.10-2.50	16	6	5	2	1	2	—
2.51-5.00	6	2	2	2	—	—	—
5.00-7.501	2	—	1	1	—	—	—
7.51-10.00	1	—	—	1	—	—	—
10.00+	1	—	1	—	—	—	—
Total	60	13	10	6	1	18	12

Source: Primary Survey, 2000

Table 12.3(*a*): Annual Income Pattern of the Sample Families in Kalyanpur Village in Midnapore District. 1999-2000

Land Holding Groups	*No. of Family*	*Annual Income of Head of the Family*	*Annual Income of other Members of the Family*	*Total Annual Income of the Families*	*Per Family Annual Income*
0.00 acres	10	188000	147000	335000	33500
0.02-2.50 acres	17	378500	403500	782000	46000
2.51-5.00 acres	18	812600	703600	1516200	84233
5.01-7.50 acres	9	751000	260000	1011000	112333
7.51-10.00 acres	4	249000	370000	619000	154750
10.00+ acres	2	72000	410000	482000	241000
Total	60	2451100	2294100	4745200	79087

Table 12.3(*b*): Annual Income Pattern of the Sample Families in Gurepara Village in Bankura District, 1999-2000.

(*in Rs.*)

Land Holding Groups	*No. of Household*	*Income of the Head*	*Income of the other members*	*Total Income of the Families*	*Per Family Annual Income*	*Per person Income*
0.00 acre	7	32500 (6)	72000 (16)	104500 (2)	14928.57	2903
0.01-1.00	27	26300 (27)	217100 (44)	480100 (71)	17781.48	3244
1.10-2.50	16	449100 (16)	339000 (17)	788100 (33)	49256.25	8566
2.51-5.00	6	322000 (6)	270000 (9)	592000 (15)	98666.66	11608
5.00-7.501	2	125000 (2)	135000 (4)	26000 (6)	130000.00	13684
7.51-10.00	1	125000 (1)	120000 (1)	245000 (2)	245000.00	30425
10.00+	1	25000 (1)	65000 (2)	90000 (3)	90000.00	11250
Total	60	1341600	1218100	2559700	42662	7071

Source: Primary Survey, 2000. Figures in the parentheses indicate the number of persons engaged in income earning.

Table 12.4(*a*): Voluntary Contribution of the Sample Families in Kalyanpur Village in Midnapore District, 2000.

Category of Families (in acres)	No. of Families	Voluntary Contribution to (in Rupees)							
		GP Tax	Political Parties	Religious Organisation	Religious Ceremonies	Sports	Voluntary Organisation	Infrastructure Development	Total
0.00	10	1.40	86.40	36.10	162.90	37.50	5.00	11.50	340.80
0.01-2.50	17	3.76	137.71	173.06	416.77	89.41	2.94	66.47	890.12
2.51-5.00	18	13.00	146.39	294.89	632.44	124.83	24.17	212.00	1447.72
5.01-7.50	9	17.10	347.89	501.33	922.67	208.44	61.11	294.89	2353.44
7.51-10.00	4	24.75	66.25	319.00	847.50	153.75	50.00	62.50	1523.25
10+	2	13.50	1636	750.50	6602.50	383.00	0.00	711.00	10096.00
Total	60	9.87	208.47	265.00	749.95	123.32	21.42	156.45	1768.67
Percentage		0.55	11.67	14.83	41.97	6.90	1.20	8.76	100

Table 12.4(*b*): Voluntary Contribution of the Sample Families in Gurepara Village in Bankura District, 2000.

Category of Families (in acres)	No. of Families	Voluntary Contribution to (in Rupees)						
		GP Tax	Religious Organisation	Religious Ceremonies	Political Parties	Infrastructure Development	Relief Work	Total
0.00	7	0.00	7.14	14.43	23.57	4.28	100.00	145.14
0.01-1.00	27	2.22	18.15	36.15	22.78	6.59	119.80	194.90
0.01-2.50	16	0.00	50.19	283.62	59.10	14.69	205.94	598.65
251-5.00	6	106.67	25.27	148.50	52.50	4.17	201.67	534.21
5.01-7.50	2	0.00	0.00	169.50	60.00	50.00	300.00	525.50
751-10.00	1	0.00	100.00	525.00	55.00	100.00	270.00	945.00
10+	1	0.00	0.00	70.00	55.00	25.00	500.00	625.00
Total	60	11.68	26.57	124.00	47.19	11.55	164.00	284.99
Percentage		3.03	6.90	32.21	12.26	3.00	42.60	100

Religious performance and religious ceremonies are important part of rural societies and rural life. In the secular states governments do not undertake any responsibility to provide and promote religious activities. The people themselves do undertake these activities. Table 12.4(a) and 12.4(b) both show that all types of village families do contribute towards religious activities in a significant way. In Kalyanpur, contribution of the village families towards these activities accounted for 56.80 per cent of their total contribution. In Gurepara this accounted for 39.11 per cent of their total contribution. In absolute terms, these were nearly Rs. 1015 in Kalyanpur and nearly Rs. 141 in Gurepara village annually.

Table 12.4(a) and 12.4(b) further show that the families of both the sample villages have contributed to the creation and maintenance of rural infrastructure. All types of families in these villages contributed labour, material and money to supplement the panchayat's scheme of constructions of roads, band (water harvesting structure) and primary school building and repairing. In money terms the annual contributions towards these activities of the families of Kalyanpur and Gurepara village were Rs. 156.45 and Rs. 11.55 respectively. These were 8.76 and 3.00 per cent of the total contribution made by an average family in Kalyanpur and Gurepara respectively.

Families in Gurepara village contributed a significant amount towards relief works. On an average, the families of the village contributed towards these works Rs. 164 and this is nearly 43 per cent of their total contribution. But this item has not been taken into account in the survey of Kalyanpur village. Village families in Kalyanpur contributed, on an average, Rs. 123.32 for undertaking sports activities in their area and this is nearly 7 per cent of their total contribution.

Another important area where the village families make contributions is the area of political activities. On an average, village families in Kalyanpur and Gurepara contributed Rs. 208.17 and Rs. 47.19 respectively for political activities in their areas. These are 11.67 and 12.26 per cent of their total annual voluntary contributions. It is worth noting that though all the development activities in a country are politically controlled by the ruling political parties in the government, no government undertakes any responsibility for developing the political consciousness of the people. Raising the political consciousness of the people still depends on the voluntary involvement and contribution of the people.

Thus, it was observed that the families in these two sample villages have voluntarily contributed a significant amount of their paltry income for undertaking various activities which, they felt, are necessary for their social life.

On the other side, these families, on an average, have paid Gram Panchayat tax to the extent of only Rs. 9.87 and Rs. 11.68 in Kalyanpur and Gurepara village. They are only 0.55 and 3.03 per cent of their total contribution respectively. These families are, however, paying a very meager amount as gram panchayat tax. People have the feeling that the state is collecting taxes for undertaking development activities which in many cases are not of their direct concern. As a result, they become hostile to pay taxes and consider taxes as a repressive measure on the part of the state power. But the same people do undertake several developmental activities which they consider as essential for their living in a 'Civil Society' as per their own conception about it, and for undertaking these activities in their area they do voluntarily contribute significant amount of money, time and energy.

These empirical findings have a very important implication for rural development. Very recently specially after 73rd and 74th Constitutional Amendments, the Gram Panchayats have been empowered to prepare and implement their own area plan. For implementing the plan there are various sources of fund—central and state government, other outside agencies, financial institutions and panchayat. To these sources it is now essential to add 'resources from people of the area' or resources from 'voluntary sector'. Local people through their contribution in terms of time, energy, money and physical resources can implement a good part of the local area plan. Instead of imposing tax by the panchayat, gram panchayat can identify the specific schemes and programmes to be implemented fully and/or partly by the local people through their contribution in terms of time, resource and money. Since the benefits of these activities will directly accrue to them, it is expected that the rural people will contribute to these activities voluntarily. Very recently works have started in this direction in a limited way. Consider the National Programme for Drinking Water Mission. Here the people of the locality are asked to contribute 10 per cent of the project cost for construction of drinking water infrastructure and they were also given the total responsibility of maintenance and operation at their own cost. Experiments along this line are being conducted in irrigation infrastructure building.

The results of these experiments, conducted on a small-scale in certain blocks in Midnapore district, are very encouraging. The feeling that the works they are doing benefit them directly motivates them to work sincerely and seriously and to voluntarily pay in terms of money and labour for carrying out these works. Thus, this approach towards rural development appears to be highly rewarding which is not so when emphasis is put on devising new taxes, imposing these taxes and collecting them for doing the works that have little effect on their standard of living, at least as the villagers feel. It is, therefore, desirable that 'voluntary sector' should be considered as an important source of fund, in addition to other sources while preparing and implementing village plan.

13

Current Thinking and Experiments in Rural Development Planning in West Bengal

Including the Convergent Community Action Strategy

RANJIT KUMAR MAITI

Introduction

Decentralized Planning is an approach to bring the planning process closer to the people so that they can decide their development priorities and participate in planning and implementation. The Panchayati Raj system provides an institutional framework in which the people play a major role in their own development. For the rural development such micro-level planning is a technique which helps in identifying developmental needs of the area and the people, prioritising them and formulating viable projects so that with limited resources maximum development could be achieved in a stipulated time period. This is derived from the alternative view of development represented by converging trends in grassroots movements which rejects the notion that development can be 'delivered' from 'above'.

The Context and the Imperatives

With the enactment of the 73rd Constitutional Amendment for Panchayati Raj System, though the three tiers of PR have their own importance, the real test of the effectiveness of the system, both in governance and development, is at the village level as the whole system should ultimately affect the life style of the people living in the villages. Under the new dispensation, the Gram Panchayats have been promised with both the requisite authority (function, functionaries) and the necessary financial resources to take the responsibility of development. Till now, almost all the development functions in the village were carried out by the government machinery and there was practically no opportunity for the village people to participate in it excepting attending Gram Sansad meetings held in May and November. Even in West Bengal the Gram Panchayats being virtually totally dependent on the financial resources from State or Central Governments have developed a dependency syndrome in which government officials have developed a feeling that they are the masters and dictate the terms to the people as to what is needed by them and what is good for them. Under these circumstances it is a challenging task to reverse the situation in which people are supposed to take the role of decision makers with the Government machinery as a helping hand to them.

Challenge of changes

Decentralization tends to disturb the existing equation and brings about new equations in the power structure betwen the elected functionaries at different layers, on the one hand, and the appointed executives, on the other.

Decentralization also affects the relations between the different tiers of development administration—both vertical and horizontal. Political power, financial allocation and the images of development functionaries and elected members are now in a vortex of change.

Imperatives of Local Level Planning

Besides these, there are other imperatives for decentralization. The implementation of the new economic reforms has indirectly meant reduced state level assistance to local areas and hence the necessity to opt for increasingly self-reliant approaches in local level planning.

The environmental crisis being the outcome of economic development, structural crisis has become more significant and is now gaining greater attention, calling for systematic environmental management and sustainable development at the local level and above.

All the above stated imperatives call for a re-orientation of the objectives, process and institutional structure of local level planning.

Implications of Decentralized Planning

In the current context the decentralized planning implies the parallel and concurrent march of democracy and decentralized development administration. It must, therefore, address itself to both the managerial as well as the political objectives.

The managerial approach involves the practical aspects of setting up appropriate planning and co-ordination mechanisms, working out organizational procedures, training and re-training of personnel involved in decentralized planning and implementation.

The political objectives are generally related to public participation usually expressed in terms of various phrases, such as 'bringing power to the people' 'involving grassroots' or 'bringing government closer to the people'.

Thus, in the context of the Panchayati Raj system, decentralisation planning implies that the managerial/technocratic approach and the political approach should mesh in an interactive, participatory and culturally sensitive planning process to produce a plan that is both technically sound and administratively responsive to the needs, wishes and resources of the local population.

Multi-levels in "Local Level"

The gram Panchayats, Panchayat Samiti and Zilla Parishad—usually called 'Local Level', form a multi-level system. Planning for such a system involves adaptation of multi-level planning procedures and effecting suitable micro-macro linkages so as to integrate local level planning into the state and national levels of planning.

It demands repetitive process of interaction and continuous convergence. This is achieved through vigorous flows of information pertaining to policies, priorities, targets, guidelines, budgetary information, technical information, co-ordination directives, review of monitoring and evaluation of projects and suggestions for re-planning.

Bottom-up flows may consist of data, information on local needs and preferences, area priorities, area targets and problems and constraints faced at the local level.

In this way the goal conflicts in multi-level planning are sought to be resolved and harmonized planning ensured. This interactive adjustment mechanism would consist of an appropriate planning machinery at each level. It is this apparatus which has to prepare the plans and organize the flows and interactions, both upward and downward. Upon the efficiency of this planning machinery depends the success of the multi-level planning.

Development from below or Development from within

Development from below is, by and large, a government initiated planning venture with a modicum of people's participation. Projects are initiated by sectoral departments and Panchayats are asked to stamp as a token of people's participation.

In the development from within the projects are basically initiated by the people. It is an endogenous development process triggered off by people themselves through trickle-up effect. It is an activity that is self-generated, self-organized and self-sustained. In this type of development, the people of a local community define their own goals, needs and aspirations and make decision effecting their well-being and sustained development. The projects that are planned by a local community with the 'untied' resources coming from above and the resources that are generated within by the local community belong to this category.

In this context of new economic reforms, the development from within holds the key to local development. It is the right tool to unleash the inner and indigenous potentials of the local community and the basis for action to realize their inner urges and performances for development through self-reliant approaches.

Development from within would make it possible for local communities to raise more resources for themselves, to assert their own value systems, to have confidence in their own ingenuity and skills and to be less dependent on external agencies for their development.

Panchayati Raj bodies must use this concept in big way to generate local resources and undertake an enlarged development effort.

This planning process is a negotiatory one in which different actors involved in the process interact, discuss and determine matters relating to resource allocation and the projects to be implemented. The matching of supply and demand constitutes the essence of decentralized planning. For arriving at the matching, the two major actors, the Government officials and the representatives of people meet and vigorously 'Negotiate' the quantum and context of the local level plan. Thus, the plan is arrived at by 'participatory consensus'. The interface takes place at GP, PS and ZP levels. It is, therefore, a process of certain top-down and bottom-up initiatives converging at the district level, resulting in a development effort that meets the people's needs with the framework of national objectives and available resources.

Planning by Gram Panchayats

According to the Eleventh Schedule of the Constitution there are 29 items which cover almost all the aspects of socio-economic life and the village people.

But the role of the PRIs at three tiers in respect of planning and implementation of the 29 items scheduled to be transferred remains vague. The various components of the 29 items have not yet been segregated and allocated to different levels of PRIs. The role and responsibilities of the PR functionaries as well as government functionaries of various departments need specific segregation. A realistic assessment of finance and authority to take investment decision at three levels require clear indication.

At present in reality in many programmes, where the Gram Panchayats have been given responsibility for implementation, they do not possess the powers of sanctioning the works and money without the concurrence of the bureaucracy. In fact, the GPs do not enjoy any flexibility in terms of tuning them to the specific needs of the villages as they are governed by the central/state guidelines.

Again with centre/state sponsored programmes another major problem is that all the activities are planned and executed within the financial year and thus it does not provide any scope for preparing a perspective plan at GP level.

Current experiments of Participatory Planning Process (under Convergent Community Action) in West Bengal

Against the above backdrop the Panchayats and Rural Development Department, Govt. of West Bengal initiated Action Research in four Pilot Blocks, namely Kashipur (Purulia), Salboni (Midnapore), Mal (Jalpaiguri) and Haringhata (Nadia) for experimentation as to see whether the Gram Panchayats can formulate their plan and implement it. The activities undertaken have been shown in Annexure A. The basis of such experiments comes from the experiences gathered from the Midnapore district where during 1982—1988 there was a massive effort for village planning by villagers.

In the present approach of such decentralised planning framework, the village is the pivotal area, where public participation can find its boldest articulation in para/tola level meetings. It is also at this level that sustainable development can be meaningfully operationalized by

Table 13.1: Planning Frame Work/Role at Different Levels

State	*District*	*Block*	*GP*
Advisory role	Sub-district level Training	GP level Training	Selection of Panchayat Resource person & local resource persons
Guiding principles	Facilitation	Facilitation	Training of PRP/LRPS
State & District level Trainings	District level planning	Manpower Mobilization	Organization of Para/Tola level meetings, special meetings with under–privileged groups.
Fund Mobilization	Mobilization of resources	Gram Sansad Meetings	Data collection in group meetings, Data analysis, problem identification, solution suggested.
Co-ordination among various Departments & Agencies	Co-ordination among various Developmental agencies	Compilation of GP level data, analysis, proritization, resource planning	Preparation of Development reports (six sectors, viz. Health, Education, women's Development, Agriculture, irrigation, animal husbandary and fisheries, Small and Cottage Industries and Infrastructure Development.
Monitoring & Review	Implementation	Technical help to GPs for project formulation	Organizing effective Gram Sansad meetings—Discussion on sectoral development reports
Effective function of State Planning & Monitoring cell	Monitoring & Review	Fund release	Compilation of data, information, suggestions from Gram Sansads
	Effective function of District level planning & monitoring cell	Implementation	Preparation of Draft GP plan (five years/one year).
		Supervision	Organizing Development seminar at GP level
		Monitoring & Review	Approval of GP plan by GP general body.
		Mid term appraisal	Final shape of GP plan, publication thereof
		Effective function of Block plg. cell	Sending proposals relating to higher tiers of panchayats Project Implementation Monitoring, Review

the people themselves, who are in daily contact with the soil, water, forests and also with each other. Thus, they are in a position to handle eco-system in a balanced way without damaging it. In terms of planning the villages may act as agencies to throw up a lot of primary data and information for planning without spending a lot of time, energy and money. They also articulate the felt needs of the people, their problems and possible solutions. These are helpful in setting proper priorities among investment proposals.

The project ideas at the village level emerge from a number of, what we may call, polycentric institutions, which are autonomous, non-hierarchical and informal in character, and in which people with similar interests can come together and contribute to solve their own problems. These are collective self-management groups-such as women's groups, beneficiary groups, joint forest management groups, user groups, rural workers' association etc. Since such groups do not automatically come into existence, a deliberate attempt is required to form such groups and for their effective functioning. Undoubtedly, leadership is required at these levels. With the state providing the necessary direction, support and protection, the local leaders and the younger generation can blossom into effective planners.

Methodology adopted in Participatory Planning Approach

- Catalytic initiatives are taken by persons coming from well-educated class, independent of macro-social organizations such as political parties, Univesity, State Training Institute to promote self-mobilization off the rural under-privileged for group or organised action to emerge out of their own deliberations.
- The starting point in generating such grass root process is to encourage the under-privileged to get together to find out why they are poor and oppressed through social investigation and analysis of their own, thus promoting their critical awareness of their environment.
- The under-privileged are encouraged to discuss what they can do jointly to overcome poverty and oppression. They are encouraged to form groups absolutely of their own, whose structure and functioning are to be decided by them, and through these groups they are required to take economic and social action according to the priorities set by them.

- Attempts are made to generate a self-reliance consciousness and an attitude to assertion of their knowledge and skill, views and decisions vis-à-vis those of outsiders. Materially, external resources and expertise are not considered to be primary in solving their problems – and these are offered only as supplements, when needed and available, for the mobilization of the people's own resources and skills. In the use of external resources emphasis is placed on the further development of people's own resources and skills to achieve progressively greater self-reliance.
- The people are encouraged to meet periodically to review their experiences, to undertake periodic fact finding investigations on their environment and to take decisions for subsequent action based on their own research thus conducted – seeking thereby to generate a process of people's practice, i.e. a progressive action – reflection rhythm.
- Dependence of the people on the initial catalysts is supposed to cease through the generation and development of internal leadership, cadres and skills. This does not necessarily mean actual physical withdrawal of the catalysts from people's process. The catalyst may continue his or her association with such process and seek new role in their progressive development.
- The initiators of such action should have not only a practical but also a research interest in generating and assisting such self-reliant people's process. This includes a search for methodology of self-reliance-sensitive catalytic action and an inquiry into the implication of such interaction for social transformation.

[Details of process and time frame followed in Action Research Projects – shown in Annexure-A]

Experience so far gained

(*a*) Those who are working with the people to promote their self-development, by and large, must not have a dogmatic position on the question of collectivism.

(*b*) The people, when they are mobilised and deliberate to set priorities and tasks, do a lot of pooling of resources and talents, and co-operation, and engage in a lot of collectivist initiatives.

(*c*) They do so as they see the objective advantage of doing so and as they feel inspired from working together to identify and solve problems and develop greater trust in each other.

(*d*) The poorer and the more oppressed the people are, the more are they likely to see, other things being equal, the advantage of such co-operation and solidarity among themselves for material improvement as well as for resisting oppression and for emotional security.

(*e*) The development of such co-operation among the people may be enhanced by sensitive 'animation' work but cannot be forced, without alienating them, by some ideological principle external to the organic evolution of their life, a principle to be applied mechanically.

(*f*) Working people could have the power to determine their own destiny within a framework of horizontal social interaction with other classes, as equals and not as inferiors.

This, ultimately, is the question of real democracy, not the democracy merely of periodic elections and the freedom to express words on what should be done, but the freedom and opportunity of the people to take the initiatives to do, to decide and to provide.

(*g*) The thinking of the people themselves may not necessarily coincide with ours, as we have serious limitations on how we define the dimensions of social development. At best, our views must be considered tentative, subject to validation or modification by the people's own process of reflection.

(*h*) A process of empowering and enabling the people to articulate and assert, by words and by deeds, their urges and thinking in this regard, must be one of the core dimensions of social development.

(*i*) Development, meaning development of people and societies, is an organic process of healthy growth and application of the creative faculties. This process may be stimulated and facilitated by external elements, but any attempt to force it towards external standards can only result in disaster.

Learning from Action Research

1. Spending about twenty five years of our early life in classrooms and studies shut off from active life to become educated and wise we have the audacity to stand above the people with our educational backgrounds and tell them how they should move, not caring even to learn from them how they have come so far struggling and moving through challenges and odds and their own thinking on issues of concern to them.
2. The educated have not proved to be any more enlightened or capable of wise and responsible decisions and conduct than the so called uneducated. While the wise persons have been responsible for the sad plight of the world today, there are numerous examples of ordinary, uneducated people devising response to problems confornting them which show great wisdom, and great sense of responsibility. The myth of vertical knowledge relation of the educated is not only factually false but also it vitally obstructs development.
3. Effective participation of the people needs time and space for discussion and for amendments of modes of thinking. Information and training should form the basis of the activities by the groups. Training should not reject traditional knowledge but will seek to strike a balance between what is positive in traditional knowledge and modern knowledge. Training and Action Research must go together to adopt new techniques, approaches as amended by the people on the basis of the progress of activities.
4. The facilitator must stimulate (*i*) a sense of self-esteem among the people; (*ii*) a sense of self-reliance on their own abilities; (*iii*) to get collectively engaged in seeking creative solution to their problems; (*iv*) mutual enrichment in a relation of equality; (*v*) solidarity and collective initiatives; and (*vi*) systematic advance.
5. The methodology of the Action Research emerges spontaneously out of the joint imaginativeness of the activists

–researchers and the people or their leaders with whom the partnership in forged.

6. Much of the activities of the people's groups reported are indeed addressed to the mechanistic notion of the development of physical assets and increasing the flow of economic and social goods and services. But the choice of meaning of development is a philosophical choice, a value judgment – as to whether the fact of the people mobilising themselves, inquiring, deciding and taking initiatives of their own to meet their felt needs, is to be regarded only as a means of development and not as an end in itself.
7. Ordinary village people in group meetings articulate on popular aspirations. The people want to stand up, take control over what they need to work with, to do things themselves in their own search for life, to move forward, supporting each other. The different articulations link with each other as if parts of the same whole. A community/society which links to move in this way, defining its tasks in favourable and unfavourable weather, becoming engaged in doing them, and reviewing their experiences to promote their self-knowledge and asserting that this is what they want to do – should not be pushed for 'development' differently so far perceived and conceptualised by us.
8. Spontaneous people's participation is not a ready-made commodity. Intervention by some external force with conscious objectives of promoting participation has yielded good result but such external intervention may create a relation of dependence of the people on him/her. Such dependence is a force against the release of people's initiatives and their collective capabilities.

Note: The views expressed are of the facilitator himself and these do not reflect views of the organization to which he belongs.

ANNEXURE A

Participatory Planning Process and Time Frame Followed (In Action Research Projects with CCA Strategy)

Sl.No.	*Activities*	*Level*	*Time Frame*
1.	Sensitisation of GP/PS/NGO/selected Panchayat Resource Persons.	Block	March, 2000
2.	Training/Facilitation of Key Resource Persons and Panchayat Resource Persons (1st Phase).	Block	April, 2000
3.	Organising Para/Tola level meeting.	Village	April-May 2000
4.	Data collection on Natural resources/Human resources/Production/Employment/Health/ Education/Infrastructure etc. – by PRPs/LRPs/GP Members.	Para/Tota/ Village	May-July, 2000
5.	Discussion on planning process in Gram Sansad meetings.	Gram Sansad	May, 2000
6.	Second Phase Training of PRPs.	Block	July, 2000
7.	Data Analysis/Identification of Problems, Resources, Solutions suggested by villagers with the help of PRPs/LRPs/GP Members.	GP	August, 2000
8.	Preparation of Gram Sansad-wise Six Sectoral Development Statements—by PRPs.	GP	October, 2000
9.	Training of Task Force Members for preparation of Draft Plan for GP (Two phases).	Block	Oct.-Nov., 2000
10.	Organizing Gram Sansad Meetings–Group discussion on Six Sectoral Development Statements.	Gram Sansad/ village	Nov.-Dec., 2000
11.	Compiliation of Gram Sansad-wise Data Common Problem/Solutions Suggested/ Proposed Schemes/Rough Cost Estimates/proposed Contribution by Villagers—By PRPs and Task Force Members [4-5 seating required].	GP	December, 2000
12.	Meeting with Block/Distirct Level departmental officials—seeking proposed flow of Departmental Fund and Target, etc. [Two Phases].	Block	November 2000
13.	Preparation of Draft GP Plan for 2001-02 with a perspective view of next 5 years – by PRPs/Task Force Members/Department Officials [Through 4-5 times meetings of the Task Force].	GP/ Block	Dec.-Jan., 2001
14.	Organizing Development Seminar for discussion on Draft GP Plan.	GP	Jan.-Feb., 2001
15.	Approval of the final GP Plan by GP's General Body.	GP	February, 2001
16.	Sending development proposals pertaining to higher tiers (PS/ZP).	GP/Block	February, 2001
17.	Publication of GP Plan-action Plan for implementation.	GP	March, 2001
18.	Preparation of Block Plan devetailing GP Plans	Block	March 2001

NOTES AND REFERENCES

1. S.P. Jain & Thomas W. Hochgesang (Eds.), *Emerging Trends in Panchayati Raj in India:* A Journey through Experience, – National Institute of Rural Development, Hyderabad (1995).
2. Rahman, Mohammad Anisur, (1994), *People's Self Development,* The Univesity Press Limited, Dhaka.
3. Tilakaratna S., (1995) The Animator in Participatory Rural Development (Concept & Practice), Geneva, ILO.
4. *Convergent Community Action – Approach Paper of* UNICEF.
5. *Rural Midnapore, An Analytical Report (1986),* Zilla Parishad Midnapore.
6. *Participatory Planning Approach Paper,* (1999-2000). SIPRD, (Govt. of WB) and Dept. of P & R.D.
7. *Process Documents on Action Research Projects in 4 Blocks* (2000-2001) – SIPRD, WB.

14

Decentralized Planning and Rural Development

An Experiment in a Block in Paschim Medinipur District of West Bengal

RUCHIRA RAHA (SARKAR) and HARIPADA JANA

Since 1985-86 attempts have been made in West Bengal to prepare micro-level plan at the block level for the development of rural people by forming Block Planning Committee at the block level, District Planning Committee and District Planning and Coordination Council at the district level. In the early 1980s, a village level plan was prepared in the district of Midnapore with active cooperation from the Rural Development Centre of I.I.T., Kharagpur. But all these attempts were sporadic, disjointed, and limited in scope. Moreover, because of the lack of people's active involvement, many of these plans could not be implemented. However, the 73rd Amendment of the Constitution has empowered the three-tier Panchayati Raj System to act as the local self-governing body, to plan for economic development, administer social justice for the people and to implement different development projects so as to ensure economic development and social justice. The said Amendment has recommended the constitution of Gram Sabha in order to ensure direct participation of the people in various development activities. This Gram Sabha, an organization of all the electorates under Gram Panchayat, can supervise and control the activities of the Gram Panchayat. Recognizing the fact that it is extremely difficult to arrive at any meaningful decisions through discussions among all the electorate-members of Gram Sabha, the West Bengal government has constituted Gram Sansad covering the area under the jurisdiction of a member of the Gram Panchayat. Thus, a Gram Panchayat will have as many Gram Sansads under its jurisdiction as the number of G.P.

members it has. For example, a 12-member Gram Panchayat will have 12 Gram Sansads under its jurisdiction. Thus, the Gram Sabha and the Gram Sansad have jointly laid the foundation of direct democracy in the rural Bengal, and have thereby facilitated the making of plans for rural development by the people for the people themselves.

This planning process will gradually replace the traditional method of preparing Annual Action Plan on the basis of allocation of funds for projects, which have been practised since long and is still being practised in the state in many Gram Panchayats. The people's plan, however, could not be implemented mainly because of the lack of funds. But relaxation of conditions associated with the Jawahar Gram Samriddhi Yojana in recent years and devolution of power to implement the District Sector Schemes to the local panchayat bodies have brightened the prospect of availability of sufficient funds. This changed situation prepared a ground for the state government to launch a movement since November, 1999 to replace the traditional planning method by the planning and implementation process at the Gram Sansad level, known as Convergent Community Action (CCA) Plan.

This paper discusses certain important aspects of the Convergent Community Action Plan (CCA) launched by the W.B. Govt. as a plan mechanism for rural development with people's participation in November 1999 in Salboni Block of Paschim Medinipur district of West Bengal under the close supervision of the Midnapore Planning and Development Society (MPDS), located in the suburb of Midnapore Town. The main objectives of the CCA Plan are (1) to sensitize the village people about the need for making various development plans by themselves; about the uses of the local resources that are available for implementing these plans; about their capability to make use of these resources and (2) provide them with technical and financial help whenever and wherever it might be considered necessary. The main thrust of the CCA Plan is that it is the people, including their elected representatives at the local level, who should play the dominant role (1) in making plans for their development keeping in mind their basic needs, hopes and aspirations and (2) in implementing these plans by using resources available from local as well as external sources.

This concept of plan starts with a conviction that people understand their problems better than any outsiders do and they know better than others what are good for them and what are not. They know more about the local resources and the use of these resources. On the other hand, their elected representatives at the higher levels or the bureaucrats, who normally frame plans for them and execute these plans, are outsiders having limited knowledge about these people, their needs and their hopes and aspirations. Therefore, they are not the right persons to make plans for uplifting the poor rural people.

To give a concrete shape to this concept of Plan, i.e., CCA, the W.B. Government started working on this plan on an experimental basis in four blocks in four different districts in the state, namely Midnapore, Purulia, Nadia and Jalpaiguri. The blocks selected for this purpose are Salboni, Kashipur, Haringhata, and Mal in these districts respectively.

The work on CCA Plan for Salboni block started under the aegis of the UNICEF and the supervision of Midnapore Planning & Development Society (MPDS). The plan started on the 24th March, 2000 by holding a sensitization camp which was attended by all elected panchayat members. The main purpose of the camp was to inform the people of the objectives of the programme. All the elected members of all the three tiers of local self government, i.e., Panchayati Raj System (PRS) and also the officers concerned of the District Administration were invited to the meeting. The meeting was followed by a two-day training camp for the Panchayat Resource Persons. The subjects discussed in that camp were

1. objectives of the CCA programme;
2. how to conduct the programme;
3. activities to be performed immediately on priority basis, and
4. the importance of convening village level or parah (a part of the village, particularly when the village is a large one) meeting.

The trained resource persons were then asked to collect different data and information on the socio-economic conditions of the people of the area for which development plans were to be made. Many questions, some of which are stated below, were raised by the people when the resource persons visited them for collecting data.

(*i*) Who would convene the parah meeting ?
(*ii*) The Panchayat members were not aware of the parah meeting.
(*iii*) The resource persons were not known to the village people.
(*iv*) How could the non-members of the Panchayats plan for development work while these works were to be performed by the Panchayat members themselves ?

(*v*) Which political parties did the Resource Persons belong to ?
(*vi*) Is it a new way of ensuring votes by suddenly showing interest and concern for the people's development just before the election?

The resource persons were thus confronted with many such questions. As a result, necessary co-operations were not forthcoming from the village people and required data and information could not be collected from the village people. Another factor that compounded this problem was that the parah meeting could not be convened in the evening since the women could hardly participate in the evening meetings mainly for safety reasons. Collection of data by the resource persons became a problem because of the large size of population from whom data were to be collected. A Gram Panchayat (G.P.) normally consists of 8 to 10 Gram Sansads and each Gram Sansad consists of at least two villages and each village has 4 to 10 parahs.(Each village is divided into 4 to 10 sections depending upon the population size of the village in order to ensure maximum attendance in the parah meeting). All these led to a situation where the resource persons, not all, were reported to have filled in the questionnaires with imaginary data. To overcome these problems, another training camp was organized where all these problems were discussed threadbare, and certain suggestions were also considered in that meeting, which were subsequently implemented. These measures were:

1. increasing the number of resource persons;
2. arranging parah meeting on those days of the month and at those times, which were convenient to both male and female persons in the villages concerned;
3. the resource persons were provided with necessary papers/ certificates as proofs of their credentials and honesty of the objectives and purpose. The resource persons were divided into different groups and each group was given an assistant selected from amongst the villagers themselves to make the collection of data smooth and easy.

In order to popularize the programme and sort out the problems facing the resource persons, there were frequent meetings of Gram (village) Sansad. In these meetings, the timings and venues for each parah meeting were fixed. These meetings helped the resource persons to acquaint themselves with the intricacies of village planning, data collection and other related issues. The concept of BPL (Below Poverty Line) was also clarified to these resource persons. Following the BPL norms, these Sansad meetings prepared a list of families lying below poverty line on the basis of the participants' knowledge about the socio-economic condition of the villagers.

Towards the end of June, 2000 a meeting of the resource persons was organized at the Midnapore Planning and Development Society Office. This meeting made an evaluation of the works of the resource persons with regard to (*i*) the experiences of working with the village people, (*ii*) the quantum and quality of responses and reactions of the participants in the village sansad meeting and of the individual respondents in respect of data/information collection from the households. Some of the reactions of the members present in Gram Sansad meetings as presented by the resource persons in that meeting are worth-mentioning here.

1. What ultimate gains could be achieved by attending the Gram Sansad meetings?
2. No points in wasting time by attending the meetings since the final decisions would be taken by only a few influential persons and political leaders and the suggestions made by the ordinary people would remain unheeded.
3. Many tall promises are made but the results have never been anywhere near these promises.
4. Development of village, in general, does not guarantee individual welfare. Some more experiences gathered from the meetings need to be highlighted here. These are:

    (*a*) only a few persons talk in the meeting and they do not allow others to ventilate their grievances or put up their view points on various issues of their interests;

    (*b*) satisfying self-interest occupies the centre-stage in the discussion. This applies to the well-to-do or socially and economically well-placed as well as the disadvantaged people equally;

    (*c*) social commitment, i.e. concern for others in the society, is missing among the members;

(*d*) attendance in the meeting is very poor;

(*e*) a section of the people in the villages, many of whom are just above the poverty line or are in a slightly better position in the society, does not want others who are poorer or not so well-placed, to come up to their levels. Fellow feeling, intentions of doing good to others, helping the people in their needs and such other qualities are missing among the villagers to a large extent.

(*f*) most of the educated people and government employees do not participate in the village sansad meeting. The retired persons hardly attend the meetings where the inexperienced common people could use their experience and wisdom for planning and executing the development programmes intended for common benefit;

(*g*) the participation rate of the women, particularly of those belonging to relatively richer and well-placed families, is very low. These women are found to be occupied more with their own family interests than with the common welfare;

(*h*) as the meetings are held at the day time, many people find it difficult to attend the meeting.

Problems Relating to Selection of Development Projects

While selecting a project as per the conventional planning, due attention is hardly is paid to the problems that are likely to emerge at the time of implementation or after the implementation since here the objective is to somehow use the fund, not so much to deliver the goods for the common benefit.

As an example of this weakness of the conventional planning, a project of digging a pond was approved by the appropriate authority without caring to know whether it would contain water during the period when water would be urgently needed for cultivation purpose. After the implementation of the project, it was found that the pond contained water only during the rainy season and no water during the summer.

Under the CCA plan various development works were selected after considering the nature and extent of risks that were likely to occur and weighing the costs associated with meeting these risks against the expected gains from the projects.

After the selection, they were arranged according to their importance in respect of meeting the immediate needs of the people concerned. This was made possible because of the participation in these meetings of the villagers who could foresee the types of problems that a particular project might face during the implementation stage.

As regards the funding of the projects, various sources were explored. These sources were, for example, (1) help from the people themselves, who are most likely to be benefited from the project or projects, in the form of physical labour or donations, (2) financial assistance from the Gram Panchayats, (3) loans from bank, (4) government grants and (5) help from the NGOs. Emphasis has been put on the first source since it implies less dependence on outside sources and greater reliance on the people's own strength.

In order to identify the problems and arrange them in order of their importance, a plan was framed in a tabular form which showed the problem area (Col. 1), problem sub-area (Col. 2), nature of the problem (Col. 3), persons affected by the problems (Col. 4), period during which the problem occurs (Col. 5), and probable solutions (Col. 6). The information on the above were collected through discussions with the village people in the parah and sansad meetings. It can be illustrated with an example.

Problem Area	*Problem Sub-Area*	*Nature of the Problem*	*Possible Reasons for the Problem*	*Period During which the problem Occurs*	*Probable Solution (s)*
1	2	3	4	5	6
Health	Health of the village people	Stomach disorder	No. of people having no pucca latrine, and no access to pipe water	During rainy season	#

Probable Solutions

(*i*) Water should be properly boiled before drinking.
(*ii*) Stopping defecation in the open field.

(*iii*) Constructing low cost latrine.
(*iv*) Educating the people about health.
(*v*) Use and supply of ORS (Oral Rehydration Solution)
(*vi*) Government facilities in this respect are to be fully utilized and so on.

The elements in the solution set were to be arranged in priority order in terms of urgency, cost and effectiveness. The solution set was divided into two sub-sets—one consisting of those solutions that could be implemented by the people themselves through their concerted efforts and the other one consisting of those solutions implementations of which required external help—financial, technical, etc. Immediate actions were to be taken to implement the suggested methods included in the first sub-set.

The report prepared on the basis of discussion in the parah meeting was placed in the village sansad meeting convened on emergency basis, wherever it was so necessary, or before the stipulated month, i.e., November, 2000, for its ratification by the sansad. However, before placing this report, the resource persons prepared another report which was read out in the sansad meeting. This report contained:

1. A general profile of social, economic and geographical features of the villages constituting the Gram Sansad and
2. the present state of the on-going projects in the village under the gram sansad.

The first part also reported the experiences of the resource persons in the preparation of the village plans regarding (1) the nature of problem encountered, (2) sources of the problem, (3) the extent of co-operation received from the panchayat members, (4) further steps to be taken for better results, and (5) the extent of impact upon the people.

As regards the geographical, social and economic features of the villages under the gram sansad, the information on the following items were collected and presented in the report:

1. Number of villages under the gram sansad;
2. Total area;
3. Total forest area;
4. Total cultivable land and proportion of irrigated area in the net cultivated area;

5. Total fallow land, total area of water bodies such as ponds, wetlands etc.; and
6. Natural resources such as river, mines, etc.

As regards the human resources, the data on the following items were collected and reported in the documents:

1. Total population;
2. Sex distribution of the total populations;
3. Number of landless labourers, marginal and small farmers, medium and big farmers.
4. Number of self employed persons
5. Caste and occupation-wise classification of the persons in the villages; and
6. Class-wise distribution of animal resources.

In the second part of the report, the current status of different projects, which are classified into six following groups, were documented. The classification of the projects is as follows:

(*i*) Education
(*ii*) Health
(*iii*) Women and children development
(*iv*) Agriculture and allied activities such as horticulture, floriculture, Pisciculture, animal husbandry etc.
(*v*) Industry and
(*vi*) Infrastructure

The assessment of present status of these projects considered the current state of progress; the problems hindering the progress of the projects, if there were any; the problems that were likely to emerge in future and the probable solutions to problems. While preparing the report on the current status of the on-going projects, care was taken to include details of each subject in the report. For example, while reporting on the current status of education in the villages under the gram sansad, the following details were mentioned.

(*i*) Number of schools—pre-primary (Anganwari), primary, junior high, high and other higher educational institutions.
(*ii*) Number of students in each institution.

(*iii*) Number of children (sex wise) (between 5 and 14) not enrolled in any school.
(*iv*) Number of drop-outs.(sex-wise).
(*v*) Number of teachers in each school.
(*vi*) Number of class rooms in each school
(*vii*) (*a*) Is there any village education committee?
(*b*) Is it functioning ? If not, why not ?
(*viii*) The present status of Iswar Chandra Jana Chetana Kendra, (ICJCK), Integrated Child Development Schemes (ICDS), Sishu Siksha Kendra, Continuing Education Centres etc.

This detailed report indicated the extent of progress (highlighting the extent of success and failure) of the project. The report was discussed threadbare in the sansad meeting and additions and alterations were made in the report on the basis of the suggestions made by the members and the report was sent to the gram panchayat office for compilation. After compilation, a draft plan for development of the gram panchayat as a whole was prepared. This draft plan was placed at a Development Seminar at the GP Level for discussion and modification. After detailed discussion in the Seminar, necessary alterations, additions and modifications were made in the Draft Plan and thus emerged the Final Plan, which was placed at the General Body meeting for final approval. The final plan consisted of the following sections:

1. *Introduction*: It contained the following information: the agricultural crop output and non-agricultural products produced; their volumes and potential given full utilization of the resources available within the area concerned; employment generated and further employment potential given the completion of the on-going employment generation programmes; the income generating government and non-government projects; the benefits accruing to the people of the area from the projects completed and underway; the standard of living of the SC and ST people; the percentage of working women and the nature of their jobs; the wage differentials, if there be any, between the male and female workers and between the workers in the relatively more advanced and backward regions or villages in the gram panchayat.

2. Geographical characteristics of the village under the gram panchayat that include quality of land, water resources, forests, mines etc.
3. Human resources available and likely to be generated, classification of human resources according to age, education, caste, religion, occupation and experience.
4. Animal resources available and likely to be created, and
5. Land use pattern and possibility of changing the pattern given the geographical characteristics and production technology.

On the basis of data and information collected on the above items, an annual plan for 2001-02 was formulated. No attempts were made to prepare a five-year-plan since there was uncertainty about the availability of fund. This annual plan presented in the tabular form consisted of the following columns: sectors and sub-sectors (Col.1&2); the plan actions (Col.3); and the labour days required for completing these activities (Col.4); materials required (Col.5); funds required for completing the project (Col. 6); the sources of fund such as the villagers themselves, the prospective beneficiaries, gram panchayat, government, banks and voluntary organizations (Cols. 7,8,9,10,&11 respectively); the prospect of employment generation in labour days (Col. 12), and wealth (Col.13); whether the projects could be done, by the gram panchayats or by other agencies and their names, (Col.14) and 'Remarks' in the last column (Col.15).

Despite all these having been done, the progress was not up to the expected level. After about six months' work, quite a good number of the resource persons were found not to have pursued the mission with the same zeal and determination as before. Some of them began harbouring such a feeling that the job entrusted to them was not their job. It was the job of the panchayat workers. Some also expressed reluctance to work further without any remuneration. However, some resource persons worked faithfully without caring for any remuneration or any favour from any quarters. They derived immense pleasure from the success of their works and were convinced that this process of 'planning by the people for the people' would deliver the goods to the people who were so long deprived of the benefits of growth of income and employment at the state or national levels. It is their determination that had ultimately paid and by the end of March, 2001 the CCA programme could be taken up in as many as nine gram panchayats out of total ten gram panchayats in the Salboni block or Panchayat Samity.

In course of implementing the CCA programme, several problems were encountered in general and also at the levels of the panchayat resource persons, government officials, and the political parties. Many of these problems were taken care of while making plans and implementing them. It may, however, be noted that if the political parties, active in the area, can be involved in the sansad level planning process, most of the problems can be solved.

A. Problems in General

1. The conventional notion about development planning is that it is the people's representatives and the administration that have the authority and expertise to make plans and implement them for people's development.
2. The people elect their representatives who will do development works on behalf of the people since they are not capable to do these difficult jobs.

B. Problems with Panchayat Resource Persons

1. All Panchayat resource persons are not equally efficient.
2. Many of them can hardly offer sufficient time for these planning exercises because of their engagement elsewhere.
3. Certain aspects such as a certain level of academic qualification, desire to work for the people and the sex should be accorded special consideration at the time of selecting resource persons, which are not very often made.
4. Lack of proper training of the resource persons poses a serious problem in the implementation of the plans.

C. Problems with the Intelligentsia and Educated Employed Persons

1. There is an inherent tendency on the part of the intelligentsia and the educated employed persons to pooh-pooh the attempts of the villagers to make plans for their development, since, according to them, the villagers do not have adequate knowledge of how to make plans.

2. These people become unnecessarily sceptical about the results of the programme much before the work starts, which quite often dampens the spirit of the villagers to prepare plans.

However, many of the people belonging to this group entertain an idea that the works are simple and they can be easily performed by the people themselves. Hence, their participation in the making of CCA plan is not needed.

D. Problems with Gram Panchayats

1. Many members of many gram panchayats are not well-educated or intelligent enough to understand the problems and the inter-linkages among various problems.
2. The methods of doing similar works vary from one gram panchayat to another depending on the level of their understanding of the problem and level of education, experience and, above all, the degree of their willingness to solve the problems.
3. Quite often, the gram panchayats do not consider this work as an indispensable part of their developmental works. The elected members, almost all, suffer from an ill-conceived notion that it is they who can do the work and, therefore, they need not discuss the matter with the people in the villages.
4. Inaccessibility to many villages or many parts of the villages in rainy season because of bad road conditions poses serious problem for the panchayat members to communicate with the villagers.
5. Many plans could not be implemented on time because of non-availability of grants from the government at the right time.

E. Problems with Villagers

1. The villagers look at these plans with scepticism since they believe the plans will never be implemented and, if at all implemented, the benefits will be reaped by the influential persons or persons belonging to the dominant political party or parties.

2. The villagers do not always supply correct information or take little interest in supplying correct information.
3. Meeting dates and times do not always suit their convenience.

F. Problems with Government Officials

1. Many of these people have a deep-rooted notion that they are educated and trained and they have enough experience in doing plans. Therefore, when the villagers, most of whom do not have the knowledge of three R' s even, embark upon doing these plans, these people look down upon them and rarely come forward with required enthusiasm and interest to help them with necessary advice.
2. Many of these officials do not consider it obligatory to participate in such exercises.
3. As many of them are already assigned various administrative works, they are left with little time and energy to attend the village/sansad or gram panchayat meetings and help the villagers to draw up development plans.

G. Problem with Political Parties

1. Co-operation from political parties had been a rare thing at the initial phase of the programme. It is, therefore, extremely necessary to seek the co-operation of all political parties from the very beginning so that no party will feel let–down or by-passed. If all the political parties that are active in the area are taken into confidence, success can be achieved without much difficulty.
2. Even when whole hearted co-operations from these parties are sought, they come in the form: 'you carry on, we are behind you, don't worry'.
3. Some times some political parties look at the planning exercises as something trivial. They give only sermons but flare up at the drop of a hat.
4. Some political parties dub this exercise as a complete hoax— just a trick to befool the people and to get the vote bank enlarged.

With sincere efforts, however, many of these problems could be tackled initially by the resource persons and later on with the help from other persons, directly or indirectly associated with the planning exercises, various development plans have been prepared for these nine G.P.s and many of these plans have already been implemented and most of the village people take pride in making substantial contributions to the socio-economic development of the villages in these GPs. It is expected that similar exercises would be performed in other GPs as well. From the Salboni experience, it can be said that the experiment has been a success and similar experiment in other GPs in the state will completely change the socio-economic scenario of the villages from the state of backwardness to that of advancement.

Acknowledgement

The authors acknowledge with thanks the help and advice received from the Prof. Purnendu S. Das, Dr. Sankar Majumdar, Mr. H. K. Samanta, the former Sabhadhipati of the district, the Sabhapati, the B.D.O. of Salboni Panchayat Samity and all the Pradhans of 10 Gram Panchayats under the Salboni Block.

15

Rural Development and Panchayati Raj Institutions

A Study Based on Household Survey of Some Villages in the District of Midnapore, West Bengal

PURNENDU SEKHAR DAS, SABARI BANDYOPADHYAY and TANIA CHATTERJEE MAJUMDER

Introduction

The Indian Government has adopted various rural development programmes that mainly aim at increasing the level of income and the rate of employment of the rural people. Worth mentioning of these programmes are Jawahar Gram Samriddhi Yojona (JGSY), Swarna Jayanti Swarojgar Yojona (SJSY), Indira Awas Yojona (IAY), Employment Assurance Scheme (EAS) and Integrated Rural Development Programme (IRDP). The Government of West Bengal is implementing these programmes through different panchayat bodies such as Gram Panchayat, Panchayat Samiti and Zilla Parishad. The panchayats in the state have been empowered by the State Government to formulate block level and district level plans and programmes since 1985-86. Every year crores of rupees are being spent on these programmes by the government through the panchayat bodies. Earlier these schemes along with various other developmental schemes were practically implemented by the state government departments through bureaucratic channels. Development plans and programmes were also formulated earlier at the state level by the Secretaries and their Assistants and approved by the Minister-in-charge of the Department concerned. As these plans and programmes made little impact on rural development, these responsibilities were later vested in the local panchayat bodies.

It is true that even after the 73rd and 74th Constitutional Amendment Acts, 1992 and 1993, which have conferred constitutional authority upon the local panchayat bodies, the status of these local bodies in respect of formulation and implementation of plans and programmes for their own betterment is far from autonomous, since most of the funds released for execution of these programmes are tied. The local bodies enjoy no freedom in doing anything outside the specified areas. It may, however, be said that in some cases the gram panchayats enjoy certain freedom, particularly in respect of selecting the village or villages and the programmes to be implemented in that village or those villages so that the local problems blocking the socio-economic development of the people and the area are solved. Again, though the release of major portion of funds in tied forms entails curtailment of autonomy, there is no denying the fact that the proper use of these funds in the government-specified programmes would certainly contribute to the development of the rural people by raising their level of income, offering employment to them, enriching the quality of their lives through mass education and primary health care programmes. Thus, setting aside the issue of full autonomy of the local panchayat bodies as an unrealized dream, we may consider how these panchayat bodies have acted on different development programmes using the powers and authority as vested in them by the state government till date in respect of implementation of various socio-economic development programmes within the limits of partial autonomy.

An assessment of the performance of the panchayat bodies can be made by studying (1) the source of receipts, particularly own receipts, (2) pattern of expenditure, and (3) utilization ratio (the ratio of total expenditure to total receipts) and ratio of each sub-group expenditure to each corresponding sub-group receipts.

This type of study gives a first hand impression about the level of performance of the local panchayat bodies regarding their power and capability of mobilizing resources and use of these resources in both financial and physical terms. But the real test of performance of a particular panchayat body lies in the evaluation of socio-economic and cultural development of the people living in the panchayat in terms of their income, employment and quality of life. Quality of life may be measured mainly in terms of their educational levels, health conditions and consumption pattern.

In this paper an attempt has been made to assess the state of socio-economic development of the people living in villages in the district of Midnapore in West Bengal and relate this state to the development expenditure made by the panchayats concerned.

Sample Design, Data Collection and Methodology

This study is based on both secondary and primary data. The secondary data pertain to receipts and expenditures of the Gram Panchayats, and these data are collected from the office records of the Gram Panchayats under study. The primary data were collected from the sample households.

The sampling design for the collection of data from the households of our study is purposive random sampling.

At the first stage, seven gram panchayats (GPs) were randomly selected from the district. These Gram Panchayats are Manoharpur, Bonpura, Baghasty, Khanamohon, Haur, Satyapur and Panchkhuri which have different socio-economic and geophysical characteristics. At the second stage, seven villages were selected, one each from these Gram Panchayats, from the list of villages belonging to the GP under study. These villages are Shyamsunderpur, Jamtola, Benadiha, Dhubni, Raisanda, Itai and Keshabpur selected in this order from the GPs, namely Manoharpur, Bonpura, Baghasty, Khanamohon, Haur, Satyapur and Panchkhuri. Once the villages were selected, samples of households were drawn at random from each of these sample villages. The sample size was proportional to the size of population and it is approximately 30 per cent of the total number of households of the villages concerned.

The number of households selected at random from these seven villages in the district of Midnapore for our study is 65, 35, 50, 30, 50, 50 and 30 respectively. The data collected through personal interview from the field survey in the villages of these districts pertain to various social and economic characteristics of the households such as land holding size, occupation structure, number of earning persons in the family, education level, enrolment of children in schools, health conditions and the like. All these data were collected during January-March of 2001. The household data refer to 2000-01. The Panchayat level data have been collected for the period from 1990-91 to 1998-99.

Analysis of Data and Results

ECONOMIC CONDITION

Land Ownership Pattern

In this section we are analyzing the economic condition of the sample households as reflected in their land ownership pattern, number of earning persons in the sample households, their economic status and occupational pattern.

Table 15.1 presents the land ownership pattern of the sample households of seven villages belonging to seven gram panchayats in the district of Midnapore. It is seen that in the village Shyamsunderpur, 5 per cent of the sample households was landless, where as the number of landless households in the villages of Jamtola, Dhubni and Itai was 3 per cent, 3.33 per cent and 6.0 per cent respectively. In Keshabpur village there was no landless household. In two villages, namely Benadiha and Raisanda, as many as 28 per cent and 26 per cent of the sample households respectively were landless.

The marginal farmers owning and/or operating less than 2.5 acres of land accounted for as much as 96 per cent of the sample households in the village Raisanda, 83.33 per cent in Keshabpur, 82 per cent in Itai, 74.4 per cent in Jamtola, 64.6 per cent in Shyamsunderpur, 56.66 per cent in Dhubni and 52 per cent in Benadiha.

Thus, the village-wise analysis reveals the following land ownership pattern of the sample households:

1. At least 52 per cent of the total sample households was landless and marginal farmers.
2. In majority of the villages considered for our study, i.e. in 4 out of 7 villages (57%), more than 70 per cent of the households was landless and marginal farmers. (in Keshabpur village there was no landless household).
3. In two villages around 80 per cent and in one village 96 per cent of the sample households were landless and marginal farmers.

Considering all the seven sample villages together it is seen that as much as 73 per cent of the total 310 sample households belonged to the groups of landless and marginal farmers, and 12 per cent were

Table 15.1: Land Ownership Pattern of the Sample Households (HH) in Seven Sample Villages

Classification of HH According to Landholding Size	*Number of Households in Sample Villages*							
	Shyamsunderpur		*Jamtola*		*Benadiha*		*Dhubni*	
	No. of HH	*C.F.<*	*No. of HH*	*C.F.<*	*No. of HH*	*C.F.<*	*No. of HH*	*C.F.<*
Landless	5 (7.7)	7.7	1 (3.0)	3.0	14 (28.0)	28.0	1 (3.33)	3.33
Less than 1 acre	15 (23.1)	30.8	14 (40.0)	43.0	9 (18.0)	46.0	10 (33.33)	36.66
1 to less than 2.5 acre	22 (33.8)	64.6	11 (31.4)	74.4	3 (6.0)	52.0	6 (20.0)	56.66
2.5 to less than 5 acre	20 (30.8)	95.4	7 (20.0)	94.4	12 (24.0)	76.0	5 (16.66)	73.32
5 to less than 10 acre	3 (4.6)	100.0	2 (5.6)	100.0	9 (18.0)	94.0	8 (26.68)	100.00
10 acres and above	0 (0.0)	0.0	0 (0.0)	100.0	3 (6.0)	100.0	0 (0.0)	0.00
Total	65 (100.0)		35 (100.0)		50 (100.0)		30 (100.0)	

Table 15.1 (*Contd.*)

Classification of HH According to Landholding Size	*Number of Households in Sample Villages*							
	Raisanda		*Itai*		*Keshabpur*		*Total*	
	No. of HH	*C.F.<*	*No. of HH*	*C.F.<*	*No. of HH*	*C.F.<*	*No. of HH*	*C.F.<*
Landless	13 (26.0)	26.0	3 (6.0)	6.0	0 (0.0)	0.00	37 (11.93)	11.93
Less than 1 acre	15 (30.0)	56.0	16 (32.0)	38.0	13 (43.33)	43.33	92 (29.68)	41.61
1 to less than 2.5 acre	20 (40.0)	96.0	22 (44.0)	82.0	12 (40.0)	83.33	96 (30.98)	72.59
2.5 to less than 5 acre	2 (4.0)	100.0	5 (10.0)	92.0	5 (16.67)	100.00	56 (18.07)	90.66
5 to less than 10 acre			4 (8.0)	100.0			26 (8.38)	99.04
10 acres and above			0 (0.0)	100.0			3 (0.96)	100.00
Total	50 (100.0)		50 (100.0)		30 (100.0)		310 (100.0)	

Note: Figures in the parenthesis are percentages.

landless. It is also found that around 42 per cent of the households includes landless and those owning less than one acre. It is distressing to note that in two out of the seven sample villages the percentages of landless households were as much as 26.0 and 28.0 with the mean of 27.0. This finding is a poor reflection on the performance of the State government and also the gram panchayats concerned in respect of distribution of excess land among the landless households. It has been one of the most important objectives of the Rural Development Department of West Bengal government to distribute through panchayats excess land among the landless and poor farmers in order to provide them with some permanent sources of income. Our findings show that this objective has remained unfulfilled even after twenty-six years' rule by this government.

OCCUPATIONAL PATTERN OF THE SAMPLE HOUSEHOLDS

Occupational pattern of the rural households indicates the state and pattern of rural development. Any change in it will mean diversification of economic activities and opening up of greater scope of earning and employment. Let us now analyse the occupational pattern of sample households of the seven villages of the district. Table 15.2 presents the information on the occupational pattern of the sample households. From the table we see that for 70 per cent of total sample households in Dhubni, 62 per cent in Raisanda, 49 per cent in Jamtola, 40 per cent in Keshabpur, 32 per cent in Itai, 30 per cent in Benadiha and 23 per cent in Shyamsunderpur, agriculture has been the main occupation. It is also seen that 20 per cent to 34 per cent of the sample households earns their livelihood as agricultural labourers in four out of seven villages, i.e. in 57 per cent of the sample villages.

Thus, agriculture and agricultural labour taken together are the main or the only source of income for about 31 per cent of the sample households in Shyamsunderpur, as many as 71.5 per cent in Jamtola, 44 per cent in Benadiha, 73.3 per cent in Dhubni, 82 per cent in Raisanda, 66 per cent in Itai and 70 per cent in Keshabpur. For all the sample villages these two are the principal occupations of as many as 60 per cent of the total 310 households. This means that the majority of the households still depend on agriculture, both directly and indirectly, to earn their livelihood. Only 40 per cent of the total sample households earn their livelihood from other sources of which service in government schools, offices or in private organizations is a major

Table 15.2: Occupational Pattern of the Sample Households

Main Occuption	*Number of Households in the Villages*							
	Shyamsunderpur	*Jamtola*	*Benadiha*	*Dhubni*	*Raisanda*	*Itai*	*Keshabpur*	*All Village*
Agriculture	15 (23.0)	17 (48.6)	15 (30)	21 (70)	31 (62)	16 (32)	12 (40)	127 (41)
Agri. Labour	5 (7.6)	8 (22.9)	7 (14)	1 (3.3)	10 (20)	17 (34)	9 (30)	57 (18.4)
Trade/Business	20 (30.7)	0	8 (16.0)					28 (9.0)
Industry	2 (3)	6 (17.1)	8 (16.0)	1 (3.3)	6 (12)	2 (4)		25 (8.1)
Service	15 (23.0)	4 (11.4)	7 (14.0)	2 (6.7)	3 (6.0)	14 (28.0)	4 (13.3)	49 (15.8)
Fisheries	3 (4.6)						0 (0)	3 (1.0)
Others	5 (7.6)		5 (10.0)	5 (16.7)		1 (2.0)	5 (16.7)	21 (6.7)
Total	65 (100)	35 (100)	50 (100)	30 (100)	50 (100)	50 (100)	30 (100)	310 (100)

Note: Figures in the parenthesis are percentages.

source of livelihood for as many as 16 per cent of the total sample households.

Thus, the analysis of the data on occupational pattern of 310 sample households spread over seven villages reveals that (*i*) agriculture still remains the mainstay as well as the main source of employment of the rural households; (*ii*) there have been no diversification of the economic activities in the rural areas and; (*iii*) no significant increase in the non-agricultural activities such as poultry farming, animal husbandry etc, which could have been easily implemented under various employment and income generation programmes.

NUMBER OF THE EARNING PERSONS IN THE SAMPLE HOUSEHOLDS

An analysis of the number of earning persons among the sample households and sex-wise distribution of the earning persons will give us an idea about the employment pattern between male and female, which will throw ample light on the status of women in the rural society, particularly in respect of much respected and used catch-word of women's independence or emancipation. Table 15.3 presents sex-wise and age-wise distribution of earning persons in the 310 sample households.

Number of earning persons belonging to the age group of 5 to 14 years is 30 only out of total 402 children in these seven villages, which is only 7.46 per cent of the total number of children. Though this percentage is very small, still this figure indicates that there is still child labour in the age group of 5 to 14 years in these sample villages, working at different places to earn for himself or herself and for the family as well.

The sex-wise distribution of the earning persons in the sample villages taken together shows that while 455 male persons out of total 624 males persons, i.e. 73 per cent are earning, only 156 out of 558 female persons, i.e. only 28 per cent earn something.

The village-wise analysis of total earning persons vis-à-vis total non-earning persons, however, presents varying pictures both at the aggregative and sex-wise levels. For example, in Shyamsunderpur, only 122 persons out of 274 persons of the sample households, i.e. 44.5 per cent earned something. This means, as much as 55.5 per cent did not earn anything or remained unemployed. Sex-wise, 115 male persons out of 144, i.e. 80 per cent and only 7 female out of 130, i.e. only 5.4 per cent earned from some sources.

Table 15.3: Sex and Age-wise Distribution of Earning Persons in the Sample Households

Village →	*Shyamsunderpur (No. of earning persons)*						*Jamtola (No. of earning persons)*						*Benadiha (No. of earning persons)*					
Age Group (year) ↓	*Male*		*Female*		*Total*		*Male*		*Female*		*Total*		*Male*		*Female*		*Total*	
	No.	*%*	*No.*	*%*	*No.*	*%*	*No.*	*%*	*No.*	*%*	*No.*	*%*	*No.*	*%*	*No.*	*%*	*No.*	*%*
5-14	2 (33)	6.06	0 (29)	0.00	2 (62)	3.22	3 (52)	5.77	0 (37)	0.00	3 (98)	3.06	0 (27)	0.00	0 (28)	0.00	0 (55)	0.00
15-60	115 (144)	79.86	7 (130)	5.38	122 (274)	44.52	51 (70)	72.86	8 (61)	13.11	59 (131)	45.04	61 (91)	67.03	8 (68)	11.76	69 (159)	43.40
Total	117 (177)	66.10	7 (159)	4.40	124 (336)	36.95	54 (122)	44.26	8 (98)	8.16	62 (220)	28.18	61 (118)	51.69	8 (96)	8.33	69 (214)	32.24

Table 15.3 (*Contd.*)

Village →	*Dhubni (No. of earning persons)*						*Raisanda (No. of earning persons)*						*Itai (No. of earning persons)*					
Age Group (year) ↓	*Male*		*Female*		*Total*		*Male*		*Female*		*Total*		*Male*		*Female*		*Total*	
	No.	*%*	*No.*	*%*	*No.*	*%*	*No.*	*%*	*No.*	*%*	*No.*	*%*	*No.*	*%*	*No.*	*%*	*No.*	*%*
5-14	– (13)	0.00	– (17)	0.00	– (30)	0.00	1 (928)	3.57	2 (29)	6.90	3 (57)	5.26	6 (28)	21.43	9 (27)	33.33	15 (55)	27.27
15-60	71 (76)	93.42	18 (66)	27.27	89 (142)	62.68	81 (97)	83.51	55 (87)	63.22	136 (184)	73.91	52 (98)	53.06	37 (90)	41.11	89 (188)	47.34
Total	71 (89)	79.78	18 (83)	21.69	89 (172)	51.74	82 (125)	65.60	57 (116)	49.14	139 (241)	57.68	58 (126)	46.03	46 (117)	39.32	104 (243)	42.80

(*Contd.*)

Table 15.3 (*Contd.*)

Village →	*Keshabpur (No. of earning persons)*						*All Sample Villages (No. of earning persons)*					
Age Group (year) ↓	*Male*		*Female*		*Total*		*Male*		*Female*		*Total*	
	No.	*%*	*No.*	*%*	*No.*	*%*	*No.*	*%*	*No.*	*%*	*No.*	*%*
5-14	4 (28)	14.29	3 (26)	11.54	7 (54)	12.96	16 (209)	7.66	14 (193)	7.25	30 (402)	7.46
15-60	24 (48)	50.00	23 (56)	41.07	47 (104)	45.19	455 (624)	72.92	156 (558)	27.96	644 (1182)	51.69
Total	28 (76)	36.84	26 (82)	31.71	54 (158)	34.18	471 (833)	56.54	170 (751)	22.64	641 (1584)	40.47

Notes:
1. Figures in the parenthesis are the total numbers of persons corresponding to the particular sex and age group in the sample households.
2. Figures in the columns of percentages are obtained from dividing the number of earning males, females and total persons in an age group by the total number of males, females and persons in that age-group.
3. Earning person means the person who earns income from some sources, such as other's land, or house, factory, office, business etc. and as labour under different rural development programmes for at least 180 mandays, and also self-employed person.

In Jamtola, 59 persons out of 131, i.e. 45 per cent were earners and therefore, 55 per cent were non-earners. Sex-wise, 51 male persons out of total 70, i.e., 73 per cent was employed and only 27 per cent did not earn anything. As regards the female persons, only 8 out of 61 could earn something, leaving 53, i.e. 87 per cent, unemployed.

In Benadiha, out of total 91 male and 68 female persons, 61 male, i.e. 67 per cent and only 8 female (12%) were earners. In other words, while among the total male adults 33 per cent were unemployed, as many as 88 per cent of the female adults were unemployed.

In Dhubni, 89 persons out of total 142, i.e. 63 per cent were earners. Of these 89 persons, 71 were male and 18 were female. Out of total 76 male persons, as many as 71 persons, i.e. 93 per cent, were earners while only 18 female out of total 66 female persons in the sample households, i.e. only 27 per cent, could earn something.

In Raisanda village, as many as 74 per cent (136 out of 184) were earners. Sex-wise, 81 male persons out of total 97, i.e. 83 per cent, and 55 out of 87 female persons, i.e. 63 per cent, were employed. Thus, Raisanda, which belongs to Haur gram panchayat, is the only village out of seven villages considered for our study where the percentage of female employment is appreciably high considering that many of the women may be either unable to work physically or unwilling to work, preferring to do non-remunerative household chores, thus helping the male or other female members in the family to earn from outside. One explanation for high employment rate for women in this village may be that the area around the village grows various kinds of vegetables and flowers for Kolkata and various suburban markets.

In village Itai, 89 persons out of total 188, (i.e. 47%) in the age group of 15 to 60 years, are earning. This means that 53 per cent of them is non-earners. Of these earners, the male earners account for 53 per cent (52 out of total 98 male persons) and female accounts for 41 per cent (37 out of 90). It is encouraging to note that this village too has a good number of women employed in various activities, though not to the extent the village of Raisanda has.

In Keshabpur village, the total number of persons in the sample households of 30 in the age group of 15 to 60 years was 104 and of them only 47 (i.e., 45%) are earning persons. Sex-wise, 41 per cent of the female persons (23 out of 56) and 50 per cent of the male persons (24 out of 48) are earners. If we make a rational assumption that some percentage of persons, whether male or female, voluntarily opts to remain non-earners, the figures of as high as 59 per cent female non-earners and 50 per cent male non-earners certainly do not give an encouraging picture of income earning employment of the people in the villages.

Now we sum up our findings from the sex-wise analysis of the earning persons in the seven sample villages in the following paragraphs:

1. In as many as five villages, the percentage of employed persons ranges between 43 per cent and 47 per cent only. This means that the percentage of unemployed persons ranges between 53 per cent to 57 per cent so far as the sample households of these villages are concerned.
2. In four villages, namely Shyamsunderpur, Jamtola, Benadiha and Dhubni, the percentage of female earning persons is very low, ranging between as low as 5.4 per cent in Shyamsunderpur to the maximum of 27.3 per cent in Dhubni. In the other three villages, this percentage is satisfactory, ranging between 41.1 per cent and 63.2 per cent.
3. As regards the percentages of male earning persons, the figures are quite satisfactory in five out of seven villages. The two villages that recorded very low percentages of male earning persons are Itai in Satyapur gram panchayat and Keshabpur in Panchkhuri-2 G.P. These two villages which form 28 per cent of the sample villages considered in our study do not show up very encouraging percentage of male earning persons either, just 41.1 per cent in each case. This phenomenon along with the prevalence of very small percentage of female earning members in as many as four villages (out of seven) bears an eloquent testimony to the failure of the panchayats to provide employment to both men and women through the implementation of various development programmes like JRY, SJSY, IAY, EAS, JGSY etc.

POVERTY LEVEL OF THE SAMPLE HOUSEHOLDS

Considering all the sample households in the seven sample villages, we find that there are as many as 69 households that are extremely poor whose poverty gap is quite large (poverty gap is given by Σgi where gi= poverty gap of the ith person = ($\prod$—Yi), where $\prod$ = poverty line, Yi is the income of the ith person). This class, thus, constitutes 22 per cent of the total 310 sample households. If we add to this the number of the poor households for whom Yi is just below $\prod$, the number of the 'extremely poor' and 'poor' households goes up to 161 and the percentage of the households lying below the poverty lines shoots up to 52 per cent. This means that more than half the number of the sample households in the randomly chosen seven villages in the district lie below the poverty line, undoubtedly a sad commentary on the performance of the panchayats which have been entrusted with the task of eradication of poverty in the rural areas, through implementation of various development programmes for which sufficient funds are allocated to each gram panchayat in the districts.

The village-wise analysis of poverty situation in the sample villages shows that except in Dhubni where the percentage of households lying below the poverty line is 33.3 per cent, which is the lowest of all, in all other villages the percentage of the households lying below the poverty line is quite high, varying between 41 per cent in Benadiha to 93 per cent in Keshabpur. Of the other villages, Jamtola has 71 per cent of the sample households and the villages—Shyamsunderpur, Benadiha, Raisanda and Itai—have between them 44 per cent to 46 per cent of the sample households below the poverty line. With respect to the percentage of extremely poor persons, i.e. the poorest of the poor, the three out of total seven sample villages, i.e. 43 per cent of the villages, have 32 per cent to 34 per cent extremely poor households. The majority of the households, however, belongs to lower middle class, which is just above the poverty line, in all the villages except Keshabpur. For all the sample villages taken together, poverty situation is really very much appalling; 22 per cent is extremely poor and about 30 per cent is poor.

The above analysis thus shows that both the mean percentage of people lying below the poverty line in these villages, as a whole, and the variations in the percentage of these people are very high.

Table 15.4: Poverty Level of Sample Households

No. of Households in the Villages		*Different Classes of Households*				
		Extremely Poor	*Poor*	*Lower middle class*	*Upper middle class*	*Total class*
Shyamsunderpur	No.	10 (15.4)	20 (30.8)	25 (38.5)	10 (15.4)	65
	CF	10 (15.4)	30 (46.1)	55 (84.6)	65 (100.0)	
Jamtola	No.	12	13	6	4	35
	CF	12 (34.0)	25 (71.0)	31 (88.6)	35 (100.0)	
Benadiha	No.	2	20	16	12	50
	CF	2 (4.0)	22 (44.0)	38 (76.0)	0 (100)	
Dhubni	No.	8	2	10	10	30
	CF	8 (26.7)	10 (33.3)	20 (66.7)	30 (100.0)	
Raisanda	No.	16	7	18	9	50
	CF	16 (32.0)	23 (46.0)	41 (82.0)	50 (100.0)	
Itai	No.	11	12	20	7	50
	CF	11 (22.0)	23 (46.0)	43 (86.0)	50 (100.0)	
Keshabpur	No.	10	18	2	-	30
	CF	10 (33.0)	28 (93.3)	30 (100.0)		
All villages	No.	69 (22.2)	92 (29.7)	97 (31.3)	52 (16.8)	310
	CF	69 (22.2)	161 (51.9)	258 (83.2)	310 (100.0)	

Note: 'Poor' people are the persons with income less than adequate for meeting 90 per cent of recommended calorie intake and 'extremely poor' are those with less than adequate income to meet 80 per cent of the recommended calorie intake. Applying these criteria involves measuring calorie requirement of the sample households spread over the district that experiences wide variation in climatic conditions, physical features and occupations across different parts of it. It also involves the problem of measuring calorie content of the food taken by the people. The food habits of these categories of people, however, include such items whose calorie and/or protein content are very low. Moreover, this definition does not consider other aspects of life, particularly the housing conditions and educational level. Therefore, we have defined 'extremely poor' and 'poor' for operational purposes in the following way and collected data on this basis.

'Extremely poor' households—the households which can not provide two full meals to any or most of the members of the household. Either all or some of the members in the family are half-fed and the food consumed by them are normally inferior quality food grains and/or vegetable. Further, all the family members live in one-room thatched house.

'Poor' households are better placed than the 'extremely poor' in the sense that these households can provide two full meals to most of the members of the family, but not to all, but they are not different in other respects from the extremely poor households.

SOCIAL CONDITIONS OF THE SAMPLE HOUSEHOLDS

The concept of development encompasses various facets of development of human beings such as social, environmental, cultural and economic development. The cultural aspects, of course, in broader sense, manifest itself, to a large extent, in the social and environmental development of the people and the area. The social development of the people can be measured by the level of education, particularly education of children and women, and health and nutrition. Let us now analyse these aspects of social development in the sample villages considered in our study.

SCHOOL ENROLMENT OF CHILDREN

Table 15.5 presents data, both in absolute value and in percentage, on the school enrolment of the children of the sample households in the seven sample villages. For all the sample villages taken together, it is observed from the table that 79 per cent of the children in the age group of 5 to 14 years is enrolled in schools and 21 per cent is not registered. If primary education is considered as compulsory for all children, then certainly the objective still remains an unfulfilled dream. This overall picture, however, conceals wide variations in the school enrolment of children among the sample villages. For example, the percentage of children not enrolled in school varies from the maximum of 40 per cent in Jamtola to the minimum of 7.3 per cent in Benadiha. This percentage is quite high, for example, 22 per cent in Itai, 20 per cent in Dhubni, 19 per cent in Raisanda and 14.5 per cent in Shyamsunderpur and 13 per cent in Keshabpur.

Sex-wise analysis of children's enrolment in schools does not show any significant percentage in the enrolment of girl students. Rather, it confirms the widely held notion that girls are discriminated against while the boys are favoured in respect of schooling. At the aggregate level, 16.8 per cent of boys and 26.0 per cent of the girls are non-registered. This means that the number of the girls not registered in school is one and a half times the number of non-registered boys. Village-wise analysis of the percentage of girls' enrolment in school shows that except in Benadiha and Keshabpur, in other five sample villages the percentage of non-registered girls in the total number of girls is very high compared to that of the non-registered boys. For example, in Shyamsunderpur, while 3 boys out of total 33, (i.e. 9%) as against 6 out of total 29 girls in the village, (i.e. 21%) is non-registered. In Jamtola, while 17 boys out of total 52 boys, (i.e. 33%) as against as many as 19 girls out total 37

Table 15.5: School Enrolment of Children

Villages	*Shyamsundarpur*			*Jamtola*			*Benadiha*			*Dhubni*		
School Enrolment Between (5 and 14 years)	*Boys*	*Girls*	*Total*	*Boys*	*Girls*	*Total*	*Boys*	*Girls*	*Total*	*Boys*	*Girls*	*Total*
Registered In Schools	30	23	53	35	18	53	25	26	51	12	12	24
	(91.0)	(79.0)	(85.5)	(67.3)	(48.6)	(59.6)	(92.6)	(92.9)	(92.7)	(35.3)	(70.6)	(80.0)
Non-Registered In Schools	3	6	9	17	19	36	2	2	4	1	5	6
	(9.0)	(21.0)	(14.5)	(32.7)	(51.4)	(40.4)	(7.4)	(7.1)	(7.3)	(7.7)	(29.4)	(20.0)
Total	33	29	62	52	37	89	27	28	55	13	17	30
	(100)	(100)	(100)	(100)	(100)	(100)	(100)	(100)	(100)	(100)	(100)	(100)

Table 15.5 (*Contd.*)

Villages	*Raisanda*			*Itai*			*Keshabpur*			*All Sample Villages*		
School Enrolment Between (5 and 14 years)	*Boys*	*Girls*	*Total*	*Boys*	*Girls*	*Total*	*Boys*	*Girls*	*Total*	*Boys*	*Girls*	*Total*
Registered In Schools	25	21	46	23	20	43	24	23	47	174	143	317
	(89.3)	(72.4)	(80.7)	(82.1)	(74.1)	(78.2)	(85.7)	(88.5)	(87.0)	(83.2)	(74.1)	(78.8)
Non-Registered In Schools	3	8	11	5	7	12	4	3	7	35	50	85
	(10.7)	(27.6)	(19.3)	(17.9)	(25.9)	(21.8)	(14.3)	(11.5)	(13.0)	(16.8)	(25.9)	(21.1)
Total	28	29	57	28	27	55	28	26	54	209	193	402
	(100)	(100)	(100)	(100)	(100)	(100)	(100)	(100)	(100)	(100)	(100)	(100)

Note: Figures in parenthesis are the percentages column-wise.

(i.e. 51%) girls have not been enrolled in any school. In Raisanda, 28 per cent of the girls as against only 11 per cent of the boys is not registered in any school. In Itai, 26 per cent of the girls as against 18 per cent of the boys have not been admitted in any school. Thus, what is observed from the above analysis is that while the mean percentage of non-enrolments for the children is not that high, the village–wise variations in this percentage is quite high and this clearly points out the glaring inadequacy in the education programmes launched in the rural areas. Further, the percentage of non-registered girls among the total non-registered children has been much higher than that of the non-registered boys in all sample villages, barring only two villages, namely Benadiha and Keshabpur (see Table 15.6 which is the reproduction of the relevant rows of Table 15.5). It is seen from the table that it is the girls who account for as much as 59 per cent of the total non-registered children in these sample villages. It has further been observed from our study that about 70 per cent to 80 per cent of the sample households with monthly income of less than Rs. 500 per month do not spend anything on their children's education, and 20 per cent to 30 per cent of the households with monthly income ranging between Rs. 1000 and Rs. 3000 spend just 2 per cent to 4 per cent of their income on their children's education.

Table 15.6: Percentages of Non-registered Girls and Boys in the Sample Villages and Households

Villages	*Non-Registered in Schools*		
	Boys	*Girls*	*Total*
Shyamsunderpur	3 (33.3)	6 (66.7)	9 (100.0)
Jamtola	17 (47.2)	19 (52.8)	36 (100.0)
Benadiha	2 (50.0)	2 (50.00)	4 (100.0)
Dhubni	1 (16.7)	5 (83.3)	6 (100.0)
Raisanda	3 (27.3)	8 (72.7)	11 (100.0)
Itai	5 (41.7)	7 (58.3)	12 (100.0)
Keshabpur	4 (57.1)	3 (42.9)	7 (100.0)
All Sample Villages	35 (41.2)	50 (58.8)	85 (100.0)

Note: Figures in the parenthesis are percentages row-wise.

EDUCATION STANDARD OF THE FAMILY MEMBERS (ABOVE 14 YEARS) OF THE SAMPLE HOUSEHOLDS

Education of the people, particularly of the women, is an important indicator of social development. The educated women play more important role in building the careers of their children than their male counterparts. Education helps to bring enlightenment, makes people aware of the evil effects of ignorance, superstition and such other things and opens up better and greater opportunities for employment. Education among women ensures, to a large extent, women's emancipation from male chauvinism. It is an investment for the present as well as for the future generations.

In the following paragraphs we study the educational standard of the family members, categorized as male and female, of the sample households of the sample villages. Table 15.7 presents data on the education standard of the households in respect of sex and level.

Considering all the seven villages together, we see from the table that 18 per cent of the people, both male and female (above 14 years) is illiterate. Of the 233 illiterate persons women account for 61.8 per cent and male persons, 38.2 per cent. [The mean difference test (difference between the mean of the percentage of male literate and the mean of the percentage of female literates in the seven sample villages) shows that the difference is statistically significant at 10 per cent probability level]. Of the total 1028 (=1261-233) literate and educated persons, 555 male persons (54%) and 473 female persons (46%) are literate or educated. Of the total 644 male persons, 500, i.e. 77.6 per cent, and of the total 617 female persons, 411 persons, i.e. 66.6 per cent, are educated at least up to primary level. From the table we observe one very interesting thing that, of the total number of persons in the sample villages educated at least up to the primary level, female persons account for 45.1 per cent, while male persons account for the remaining 54.9 per cent. [This difference between the percentages of the male and female persons for all the seven villages, is, however, found to be statistically significant at 10 per cent probability level.]

It is further observed that of the total 644 male persons, only 89 persons (14%) and 144 out of total 617 female persons (23%) are illiterate.

Considering the villages individually it is seen there is wide variation in the percentage of illiterate persons. In as many as four villages, which constitute 57 per cent of the sample villages, this percentage is very high (29.61). In the other three villages, this percentage is very low (6.82). As regards the village-to-village variations in the percentage of illiterate persons it is seen that in three out of the total seven sample villages this percentage ranges between 10.2 per cent and 13.1 per cent. In the other four villages, this

Table 15.7: Educational Standard of the Family Members of the Sample Households (Above 14 years)

Village / Education Level	Shaymsunderpur			Jamtola			Benadiha			Dhubni		
	Male	*Female*	*Total*	*Male*	*Female*	*Total*	*Male*	*Female*	*Total*	*Male*	*Female*	*Total*
Illiterate	18	20	38	19	16	35	2	0	2	0	2	2
	(11.1)	(15.9)	(13.1)	(29.69)	(19.75)	(24.14)	(2.5)	(0.0)	(1.25)	(0.0)	(1.2)	(1.2)
Literate	20	24	44	18	14	32	5	8	13	6	15	21
	(12.3)	(19.0)	(15.2)	(28.13)	(17.28)	(22.07)	(6.25)	(10.0)	(8.12)	(3.6)	(9.0)	(12.5)
Primary	85	68	153	21	38	59	22	41	63	13	21	34
	(52.5)	(53.9)	(53.1)	(32.81)	(46.92)	(40.69)	(27.50)	(51.25)	(39.38)	(7.7)	(12.5)	(20.2)
Madhyamik	15	10	25	6	9	15	32	24	56	50	37	87
	(9.2)	(7.9)	(8.6)	(9.37)	(11.11)	(10.34)	(40.0)	(30.0)	(35.0)	(30.0)	(22.0)	(51.8)
Higher Secondary	8	2	10	0	2	2	11	5	16	7	2	9
	(4.9)	(1.6)	(3.5)	(0.00	(2.47)	(1.38)	(13.75)	(6.25)	(10.00)	(4.2)	(1.2)	(5.4)
Graduate	12	2	10	0	2	2	5	2	7	4	3	7
	(7.4)	(1.6)	(3.5)	(0.0)	(2.47)	(1.38)	(6.25)	(2.50)	(4.37)	(2.4)	(1.8)	(4.2)
Post Graduate	2	0	2	0	0	0	3	0	3	0	1	3
	(1.2)	(0.0)	(0.6)	(0.0)	(0.0)	(0.0)	(3.75)	(0.0)	(1.880	(2.2)	(0.6)	(1.8)
Diploma	2	0	2	0	0	0	0	0	0	5	–	5
	(1.2)	(0.0)	(0.6)	(0.0)	(0.0)	(0.0)	(0.0)	(0.0)	(0.0)	(3.0)		(3.0)
Total	162	126	288	64	81	145	80	80	160	87	81	168
	(100.0)	(100.0)	(100.0)	(100.0)	(100.0)	(100.0)	(100.0)	(100.0)	(100.0)	(52.0)	(48.2)	(100.0)

(Contd.)

Table 15.7 (Contd.)

Village / *Education Level*	*Raisanda*			*Itai*			*Keshabpur*			*All Sample Villages*		
	Male	*Female*	*Total*	*Male*	*Female*	*Total*	*Male*	*Female*	*Total*	*Male*	*Female*	*Total*
Illiterate	10 (10.37)	31 (32.29)	41 (21.20)	19 (18.0)	40 (41.68)	59 (29.35)	21 (42.86)	35 (61.40)	56 (52.86)	89 (13.82)	144 (23.34)	233 (18.48)
Literate	2 (2.06)	1 (1.04)	3 (1.55)	4 (3.8)	0 (0.0)	4 (1.99)	–	–	–	55 (8.54)	62 (10.05)	117 (9.28)
Primary	33 (34.03)	29 (30.21)	62 (32.1)	20 (19.0)	20 (20.83)	40 (19.90)	17 (34.69)	16 (28.07)	33 (31.13)	211 (32.76)	233 (37.76)	444 (35.21)
Madhyamik	45 (46.39)	28 (29.17)	73 (37.82)	34 (12.4)	23 (23.96)	57 (28.36)	11 (22.45)	6 (10.53)	17 (16.04)	193 (29.97)	137 (22.21)	330 (26.17)
Higher Secondary	3 (3.09)	4 (4.17)	7 (3.64)	16 (15.2)	8 (8.3)	24 (11.94)	–	–	–	45 (6.99)	23 (3.73)	68 (5.39)
Graduate	1 (1.03)	3 (3.12)	4 (2.07)	10 (9.5)	3 (3.12)	13 (6.47)	–	–	–	32 (4.97)	15 (2.43)	47 (3.73)
Post Graduate	2 (2.06)	0 (0.00)	2 (1.04)	2 (1.9)	1 (1.04)	3 (1.49)	–	–	–	11 (1.71)	2 (0.32)	13 (1.03)
Diploma	1 (1.03)	0 (0.0)	1 (0.52)	0 (0.0)	1 (1.04)	1 (0.50)	–	–	–	8 (1.74)	1 (0.16)	9 (0.71)
Total	97 (100.0)	96 (100.0)	193 (100.0)	105 (100.0)	96 (100.0)	201 (100.0)	49 (100.0)	57 (100.0)	106 (100.0)	644 (100.0)	617 (100.0)	1261 (100.0)

Note: Figures in the parentheses are percentages column-wise.

percentage lies between 21 per cent in Raisanda and 53 per cent in Keshabpur. Jamtola and Itai have 24 per cent and 29 per cent illiterate persons respectively. Again, among the male and female persons taken together the percentage of illiterate female persons is very high in comparison to that of the illiterate male persons in four villages, namely Raisanda (32.29, 10.31), Itai (41.68, 18.0), Keshabpur (61.40, 42.86) and Shyamsunderpur (15.9, 11.1) (The first number and the second number within bracket refer to the percentage of female illiterate and male illiterate persons respectively). In Jamtola, however, the percentage of illiterate female persons (19.75) is lower than that of male persons (29.69).

The analysis of sex-wise distribution of the total number of illiterate persons in each village reveals that in four villages the female illiteracy is very high as compared to the male illiteracy. These villages and their female illiteracy percentages are as follows:

1. Shyamsunderpur (52.63),
2. Raisanda (75.61),
3. Itai (67.80) and
4. Keshabpur (62.50).

For all the seven sample villages taken together the female illiteracy percentage in the total illiterate persons is 62 per cent.

Thus, the above analysis shows that the mean illiteracy rate in these seven sample villages is around 18 per cent, but there is wide variation in the literacy rate among the villages, ranging from 1 per cent to as high as 53 per cent. Of the illiterate persons, the female illiteracy is, in most of the cases, much higher than the male illiteracy. In four villages, i.e. in 57 per cent of the villages under study, the female illiteracy rate (among total illiterate persons) ranged between 53 per cent and 76 per cent. These data uncovered the hollowness of the claims of achieving 100 per cent literacy in the districts.

HEALTH AND SANITATION

Health condition of rural people is an important indicator of rural development. Good health and high average life expectancy taken together will imply that people get at least two square meals a day that meet the minimum calorie requirements of them. Considering life expectancy alone as a measure of economic development may be misleading because many people living in the villages with frail bodies, even sometimes without having two square meals a day and suffering from various minor diseases have been found to live long. Many of them have been found to have lived beyond 60 years. Therefore, the average life expectancy factor alone does not always convey many important aspects of physical development of people such as condition of health and standard of living. Similarly, a person with poor health, measured in terms of height and weight, as is normally done, may also live longer despite his/her poor health and a person with good health may not live long as expected. Therefore, instead of measuring health in terms of height and weight and certain other physical characteristics if one measures it in terms of the types of diseases the people normally suffer from, the sanitary facilities they enjoy, quality of drinking water, environmental pollution within and outside their houses and such other things, one can get a better measure of health condition of the people, in general. In the case of children, the extent of use of health care facilities may be used as a better measure of health.

In this section we try to determine the health condition of the rural people in terms of the following characteristics: kinds of disease normally attacking the poor in the rural area, sources of drinking water; types of latrines and the use of smokeless oven. Inoculation or non-inoculation and the number of child death in these group of 0 to 5 in a family in the last five years are considered as a measure of the health condition of the children. Tables 15.8 and 15,9 present data on health conditions of the children and adults in terms of the above mentioned factors. From Table 15.8 it is seen that under–five mortality rate [q(5)] widely varied among the villages, ranging from as high as 273 children per thousand live births in Benadiha village to only 48 in Dhubni. But in three out of seven villages this number was very high (viz. 273,195,197); in one village it was 133 and in three other villages it was 95, 48 and 42 only. However, the mean under–five mortality rate was as high as 140 per 1000 live births in the study year. This figure may be compared with the 1981 figure for rural West Bengal which was 139. In the last twenty years there should have been appreciable decline in this mortality rate in West Bengal. But our data show that even in 2001 there are many villages where this rate is much higher than even the 1981 figures.

Table 15.8: Health Care Arrangement for Children and Family

Villages	*Benadiha*						*Raisanda*						*Shyamsunderpur*						*Jamtola*					
Health Condition	*Very very poor*	*Very poor*	*Poor*	*Lower middle class*	*Upper middle class*	*Total HH*	*Very very poor*	*Very poor*	*Poor*	*Lower middle class*	*Upper middle class*	*Total HH*	*Very very poor*	*Very poor*	*Poor*	*Lower middle class*	*Upper middle class*	*Total HH*	*Very very poor*	*Very poor*	*Poor*	*Lower middle class*	*Upper middle class*	*Total HH*
Total no. of HH	0	2	20	16	12	50	13	3	7	18	9	50	2	8	20	25	10	65 (100)	5	7	13	6	4	35 (100)
Mortality rate per 1000 live (q5)[1]	-	-	-	-	-	273	-	-	-	-	-	152	-	-	-	-	-	195	-	-	-	-	-	133
Non-inoculated children	-	-	-	-	-	-	-	-	-	2	1	3	1	0	0	0	0	1	0	0	0	0	0	0
Smoke-less oven (no. of HH)	0	0	1	5	7	13 (26)	0	0	0	2	1	3 (6)	0	0	0	10	8	18 (27.69)	0	0	1	2	3	6 (17.14)
Latrine (no. of HH)	0	0	1	7	10	18 (36)	3	1	4	9	9	26 (52)	0	0	0	15	10	25 (38.36)	0	0	1	3	3	7 (20)
Drinking water source (no. of HH)	Tube wells						Tube wells						Tube wells						Tube wells					

(Contd.)

Table 15.8 (*Contd.*)

Villages	*Dhubni*						*Itai*						*Keshabpur*						*All Sample Villages*					
Health Condition	*Very very poor*	*Very poor*	*Poor*	*Lower middle class*	*Upper middle class*	*Total HH*	*Very very poor*	*Very poor*	*Poor*	*Lower middle class*	*Upper middle class*	*Total HH*	*Very very poor*	*Very poor*	*Poor*	*Lower middle class*	*Upper middle class*	*Total HH*	*Very very poor*	*Very poor*	*Poor*	*Lower middle class*	*Upper middle class*	*Total HH*
Total no. of HH	7	1	2	10	10	30	3	8	12	20	7	50	-	10	18	2	-	30	30	39	92	97	52	310
Mortality rate per 1000 live (q5)[1]	-	-	-	-	-	48	-	-	-	-	-	95	-	-	-	-		197	-	-	-	-	-	156
Non-inoculated children	-	-	-	-	-	-	1	1	5	0	0	7	-	2	1	-	-	3	2	4	6	2	-	14
Smoke-less oven (no. of HH)	-	-	-	3	3	6 (20)	0	0	1	2	3	6 (12)	-	0	1	-	-	1 (3.33)	-	-	4	24	25	53 (17.1)
Latrine (no. of HH)	-	-	1	7	10	18 (60)	1	3	8	15	5	32 (64)	-	-	1	1	-	2 (6.66)	4	4	16	57	47	128 (41.20)
Drinking water source (no. of HH)	Tube wells						Tube wells						Tube wells						Tube wells					

Notes: 1. Figures in the parentheses indicate percents in respect of total number of households.
2. Mortality rate (q5) is the number of deaths of children upto the age of 5 years per 1000 live births.

Table 15.9: Common Diseases of the Rural People

Ordinary Diseases (Suffered during the last year)	*Shyamsunderpur*			*Jamtola*			*Benadiha*			*Dhubni*		
	Child	*Adult*	*Total*	*Child*	*Adult*	*Total*	*Child*	*Adult*	*Total*	*Child*	*Adult*	*Total*
Ordinary stomach diseases	41 (39.80)	98 (33.22)	139 (34.92)	25 (21.0)	32 (22.07)	57 (21.59)	10 (15.15)	20 (12.50)	30 (13.27)	51 (100.0)	110 (65.48)	161 (73.52)
Fever and cold	59 (57.28)	87 (29.49)	146 (36.68)	35 (29.41)	45 (31.03)	80 (30.30)	35 (53.03)	28 (17.50)	63 (27.88)	51 (100.0)	100 (59.52)	151 (68.95)
Skin diseases	21 (20.39)	25 (8.47)	46 (11.56)	12 (10.08)	22 (15.17)	34 (12.88)	11 (16.66)	17 (10.62)	28 (12.39)	–	5 (2.98)	2 (2.28)
Other ordinary diseases	46 (44.66)	38 (12.88)	84 (21.10)	5 (4.20)	11 (7.59)	16 (6.06)	2 (3.03)	7 (4.37)	9 (3.98)	3 (5.88)	100 (59.52)	103 (47.03)
Total number of persons in the sample households	103	295	398	119	145	264	66	160	226	51	168	219

Table 15.9 (*Contd.*)

Ordinary Diseases (Suffered during the last year)	*Raisanda*			*Itai*			*All Sample Villages*		
	Child	*Adult*	*Total*	*Child*	*Adult*	*Total*	*Child*	*Adult*	*Total*
Ordinary stomach diseases	8 (9.88)	39 (20.0)	47 (17.03)	27 (32.93)	34 (16.91)	61 (21.55)	162 (32.27)	333 (28.61)	295 (17.70
Fever and cold	12 (14.81)	43 (22.05)	55 (19.93)	30 (36.58)	16 (7.96)	46 (16.25)	222 (42.22)	319 (27.40)	541 (32.47
Skin diseases	1 (1.23)	6 (3.08)	7 (2.54)	12 (14.63)	17 (8.46)	29 (10.25)	57 (11.35)	92 (7.90)	149 (8.94
Other ordinary diseases	21 (25.92)	89 (45.64)	110 (39.85)	4 (4.48)	6 (2.98)	10 (3.53)	81 (16.13)	251 (21.56)	332 (19.93
Total number of persons in the sample households	81	195	276	82	201	283	502	1164	1666

Note: Figures in the parentheses are the percentages of the corresponding numbers with respect to the total numbers shown in the last row of the table.

The number of non-inoculated children is only 14 out of total 581 children (in the age group of less than 14 years) in the sample households. Though this figure tells us that there are still some households who could not avail themselves of the child health care facilities, it is, no doubt, a no mean achievement that as many as 97 per cent of the children could be inoculated.

It is hoped that the awareness about the importance of the children's health care among majority of the households, (about 98%, i.e. 303 households out of total 310 sample households) will bring the uncovered households under the health care system very soon. The village-wise analysis shows that in three villages (out of total seven sample villages) there were no non-inoculated children and in the other four villages, there were only 1 non-inoculated out of total 103 children in Shyamsunderpur, 3 among 81 in Raisanda, 7 among 82 in Itai and 3 among 79 children in Keshabpur. The highest percentage of non-inoculated children (8.53) could be found in Itai village. Comparing this valuable information with the q(5) mortality rate (where q(1) morbidity rate has the largest share), one can safely say that along with the child health care pre-natal care of pregnant mothers should be given special attention to prevent q(1) mortality rate, (which has the largest share in q(5) mortality rate), which has not been done so far.

Thus, it can be said that the villages, in general, have made significant progress in respect of child care. It is a no mean achievement on the part of their local authorities to awaken the rural poor people to meeting their basic needs. It may be mentioned that it is the poor households, including the very very poor and very poor, which have as many as 11 non-inoculated children out of total 14 non-inoculated children, which accounts for 80 per cent of the total non-inoculated children.

Now we consider three other aspects of rural development which have significant bearings on the standard of living of the people in the rural areas. These are: the use of smokeless oven, latrine and the sources of drinking water. Only 53 households out of 310 households, (i.e. 17%) have smokeless ovens and 128 households, i.e. 41 per cent, use latrines. These results show that most of the villagers still use the traditional means of cooking their foods etc. and open space for defecation that pollutes the water bodies, contaminates the pond and well water that is, more often than not, used by many rural households as drinking water, particularly when the tube wells dry up in the summer. Defecation on the banks of the pond is a common sight in the rural areas. Most of the rural households, it has been noted during the survey, use pond water for cooking, particularly for cooking rice on the ground that rice boiled in tube well water loses its natural colour and odour and eating that rice gives a bad taste in the mouth. Advice on bad effects of the use of this type of pond water meets with the ready reply that as water is boiled, it must not have any significant bad effect upon the consumers and they drink tube well water whenever available. It may be noted here that though the households reported using tube well water for drinking purposes, most of them have to fetch water from tube wells located at distant places.

The distance as well as the frequent breakdown of the tube wells force the villagers to use pond water for drinking purposes. This, as we shall see from Table 15.8, is one of the main causes of various stomach ailments of both children and adults in the sample households. About 32.27 per cent and 28.61 per cent of the total number of children and adults in the sample households respectively suffered from ordinary type of stomach diseases and 42.22 per cent and 27.40 per cent of children and adults suffered from cold and fever during the last one year preceding the survey period. Skin diseases are also not very uncommon among the rural people. The percentage of persons suffering from skin diseases is 8.94 per cent. The use of impure water for drinking and cooking food and polluted surroundings are the main causes of this illness.

Though the percentage of different types of ailments among the children and adults varies from village to village, 'fever and cold' has been the main disease in all the six villages considered here, followed by skin diseases in terms of persons attacked. (Data on diseases could not be collected from one village – Keshabpur). It is further observed that the percentages of children attacked with stomach diseases and cold and fever are more than those of the adults in four out of six villages, i.e. in 67 per cent of the villages under study. The percentage of children suffering from skin disease s is also relatively higher in three villages. The data on diseases, thus, show that about 91 per cent of the persons, including children and adults, suffered from one or the other or all or some of these diseases during the year preceding the period of survey.

In this connection it may worthwhile pointing out that about 40 per cent of the sample households rarely take meat or chicken as one of the courses in their meals. About 20 per cent consumes either of these two non-vegetarian items once in a month or so. After the green revolution, raising and raring of cows have been declining very sharply, particularly among the poor people, because of their inability to provide them with sufficient fodder. Therefore, production and consumption of milk has become a luxury for these people who constitute about 40 per cent to 50 per cent of the sample households.

In most of the villages, namely Keshabpur, Jamtola, Shyamsunderpur, Raisanda and Itai food items consumed by most of the households rarely provide the required calorie and hence make them vulnerable to various diseases.

We now briefly set forth our findings on the social aspects of rural development in the following paragraphs:

1. (*a*) There is still high percentage of the children in the villages who are not enrolled in schools.
   (*b*) Among the children the percentage of non-enrolled girls is much higher than that of the boys in almost all villages. Sex-discrimination is still prevalent in the rural area in a very significant way.
   (*c*) Among the adult persons in the sample villages the mean percentage of illiterates is around 30 per cent and this percentage varied from slightly more than 1 per cent to as high as 53 per cent.
   (*d*) Among the illiterate persons the percentage of illiterate female persons is very high in relation to the male illiterate persons. In four villages this percentage ranged between 52.63 to 75.61 with the mean value of 64.63.
2. The use of health care facilities by the children of the sample households is quite satisfactory so far as the number of non-inoculated children is concerned.
3. (*a*) Of the total 1261 persons in the sample households in seven sample villages, 233 persons, i.e. 18.48 per cent, are illiterate and 9.28 per cent is just literate. Of the total illiterate persons as many as 144 female and only 89 male persons, i.e. 62 per cent and 38 per cent respectively, were illiterate. However, it is interesting to note that among the total literates and primary educated persons, the percentages of female persons are 53 and 52 respectively. [The differences between the percentages of female

literates and the male literates and between the percentages of primary educated female and male persons are not statistically significant even at 20 per cent probability level. (The t test has been employed to test the difference of means of two series of data on percentages of female literates and male literates respectively for seven villages. In this case $t = 0.06$. The assumption of equality of variances used in the t test has also been tested by applying F test. The value of $F=1.24$ was found to be statistically not significant)].

(*b*) As regards the other aspects relating to health care and environment it has been found that there is little use of latrine and smokeless oven in the countryside.

These findings provide some glaring examples of the failures of the Gram Panchayats in making provisions of the basic civic amenities, particularly drinking water and latrine, for the villagers.

Let us now discuss the performance of the gram panchayats under study in respect of utilization of fund received from various sources like the Government funds, both tied and untied. The data on total funds available for developmental activities from Government grants and own sources, total expenditures on development activities, utilization of funds (ratio of developmental expenditures to development receipts into 100), funds available to each Gram Sansad and each mouza for developmental activities are presented in Table 15.10 below. We also present data on some important economic and social indicators of development in the same table in order to judge the development status of the villages alongside the funds available per sansad and per mouza for development activities. So far as the utilization of fund by these GPs is concerned, it is found that during the period under study the average utilization ratio varied from the lowest 61.11 per cent in Mohanpur–1 GP to the highest 83 per cent in Bonpura G.P. The average utilization rate for all these GPs was 72.72 per cent only. In other words, these GPs failed to utilize 17 per cent to 40 per cent of the fund available to them during the period under study. One of the reasons normally cited in support of non-utilization of fund is the receipt of a sizable amount of fund from the State Governments towards the end of the financial year, normally in February and March. Another reason is the GPs' ignorance about the amount of fund that might be available in a year, which prevents these panchayats from making plans for mobilization of necessary resources for utilization of funds. There may be many other reasons, but given the above situation one has to accept non-utilization of fund to the extent of about 30 per cent as inevitable. However, inspite of this limitation, the amount that has been used by each Gram Sansad during the period under study (1990-1 to 1998-9) for development purposes is 9.75 lakhs in one G.P., over 5 lakhs in three G.Ps, about 3 lakhs in one G.P. and around 2 lakhs in two G.Ps. (serial no. 4 in the table) and the amount available in each mouza for development purposes ranged between the lowest Rs. 1,14,245 in Satyapur GP and the highest Rs. 5,86,618 in Bonpura GP during this period (see serial no. 5). The amount available per mouza was, on an average during this period, 2.87 lakhs approximately. Proper utilization of this fund might have produced much better results than what the figures presented under serial numbers 6 to 13 in the Table 15.10 indicate.

Here it may be claimed that many significant performances by the GPs such as construction of roads or school buildings and digging of tube-wells have not been considered in this study. But it can be argued that the extent of achievements and quality of performances in these areas are, no doubt, reflected in the basic development indices considered by us in this study. For example, in respect of generation of non-farm employment in rural areas, in only two villages the percentage of households engaged in non-farm activities is more than 50 and in the other five villages it ranges from 18 to 34 with the mean of 27 only. In other words, for as many as 73 per cent of the households agriculture was the only source of employment and income. If this result is assessed in the context of powers vested in the GPs by the 73rd Amendment of the Indian Constitution (1992), namely (*i*) promotion and encouragement of cottage, khadi, village and small-scale industries including food processing industries, (*ii*) promotion of dairy farming and poultry, (*iii*) promotion of fishery and (*iv*) maintenance of community assets etc., it appears that the Gram Panchayats under study have failed to take necessary steps in this direction.

As regards the employment generation of men and women in the rural areas, which is one of the main objectives of the 73rd Amendment of the Constitution, the situation does not appear to be very encouraging, particularly for the women. The percentage of female earning persons among the total adult female persons in our sample villages is very

Table 15.10: Development Receipts, Expenditures, and Some Important Development Indicators for the GPs and the Villages under Study

Funds of GPs & Development Indicators ↓ / Villages (G.P.) →	*Shyamsunderpur (Manoharpur G.P.)*	*Jamtola (Bonpura G.P.)*	*Benadiha (Baghasty G.P.)*	*Dhubni (Khana Mohan G.P.)*	*Raisanda (Haur G.P.)*	*Itai (Satyapur G.P.)*	*Keshabpur (Panchkhuri II G.P.)*	*All Villages*
(1)	(2)	(3)	(4)	(5)	(6)	(7)	(8)	(9)
1. Total fund available for development purpose[1]	48,53,251	84,65,796	1,11,03,758	1,74,59,895	59,14,60 3	70,79,056	1,04,16,587	652.9 lakhs
2. Total amount spent during the period[2]	29,65,746	70,39,412	77,29,357	1,36,55,005	48,46,33 5	50,26,789	66,66,898	479.3 lakhs
3. Utilisation of fund (%) [= (2) ÷ (1)]	61.11	83.15	69.61	78.21	81.94	71.01	64.00	73.41
4. Fund that has been used for development activities per sansad (Rs.) [2÷No of sansads]	1,97,716	5,02,815	5,52,097	9,75,357	2,42,317	2,95,693	5,55,575	474510
5. Fund that has been used for development activities per mouza (Rs.) [2÷No of mouzas]	1,64,764	5,86,618	2,41,542	2,13,359	1,79,494	1,14,245	5,12,838	287551
6 (*a*) Ratio of HHs engaged in non-agricultural activities to total sample HHs (%) (*b*) (Index)[3]	69.0 (88.00)	28.0 (6.00)	56.0 (62.00)	30.0 (10.00)	18.0 (–14.00)	34.0 (18.00)	30.0 (10.00)	40.64 (31.28)
7. Ratio of number of earning persons to the total number of persons in the sample HHs (%)	44.52	45.04	43.40	62.68	73.91	53.06	47.34	52.38

(*Contd.*)

Table 15.10 (*Contd.*)

(1)	(2)	(3)	(4)	(5)	(6)	(7)	(8)	(9)
8. Ratio of female earning persons to the total number of male earning persons in the sample HHs (%)	43.40	15.69	13.11	25.35	67.90	71.15	95.83	34.28
9. Ratio of female earning persons to the total number of female persons in the sample HHs.	5..38	13.11	11.96	27.27	63.22	41.11	41.07	26.62
10. Ratio of HHs lying APL to the total number of sample HHs (%)	53.58	28.57	56.00	66.66	54.00	54.00	6.67	48.06
11. q(5) Survival rate (per 100 live births)	80.50	86.70	72.70	95.20	84.80	90.50	80.30	87.00
12. (*a*) Ratio of combined primary, secondary and tertiary educated persons to the total number of persons (%)	86.57	69.66	97.20	95.96	79.20	72.25	60.62	80.88
(*b*) Adult literacy rate (%)	60.12	55.53	72.12	76.51	60.15	75.32	73.89	67.66
13. Education Index[4]	68.85	60.19	80.40	86.95	66.44	74.30	72.44	72.97
14. Total Development Index[5]	56.80	39.93	54.41	58.13	54.73	55.16	42.97	53.05

Notes:

1. Opening balance +Govt. Grants for Development Purposes + Own Source Receipts.
2. Opening Balance +Govt. Grants +Own source receipts—Closing Balances.
3. This Index was calculated by the following formula: $\frac{\text{Actual value} - \text{Minimum value}}{\text{Maximum value} - \text{Minimum value}} = \frac{\text{Actual value} - 25}{75\text{–}25}$
4. This Index has been calculated by the following formula: $[\,12(a) \times 0.33] + [\,12(b) \times 0.67]$
5. Total Development Index is the simple average of the development indicator (6, 7, 9, 10, 11 and 13). While calculating index for the indicators other than 6, the minimum and maximum values have been fixed at 0 and 100 respectively.

low ranging from 5.38 to 27.27 in the four out of seven sample villages with a mean of 14.43 only. In the other three villages the percentage varied from 41.07 to 63.22 with a mean of 48.47. The average of the percentage of female earning persons is estimated to be only 29, which does not indicate a satisfactory achievement by any standard.

Thus, considering the development indicators such as ratio of BPL households to the total number of sample households and under–five mortality rate (per 100 live births), one can safely conclude that the Gram Panchayats under study could not play any responsible role in altering the socio-economic scenario in the rural areas. A total development index has been computed for each sample village as well as for all the sample villages taken together on the basis of six development indicators (serial numbers 6(b), 7, 9,10, 11 and 13 in Table 15.10), according to the general formula used in the computation of Human Development Index (HDI) for any country or state or region. In all the sample villages, this index remained below 60, which in no way indicates satisfactory level of performances by the GPs.

Index